C614

Crops in construction handbook

Andrew Cripps   Buro Happold

Richard Handyside   Construction Resources

Liam Dewar   Construction Resources

Jonathan Fovargue   Construction Resources

CIRIA *sharing knowledge ■ building best practice*

CIRIA, Classic House, 174-180 Old Street, London EC1V 9BP, UK.
Telephone: +44 (0)20 7549 3300   Fax: +44 (0)20 7253 0523
Email: enquiries@ciria.org   Web: www.ciria.org

Crops in construction handbook

Cripps, A; Handyside, R; Dewar, L and Fovargue, J.

Keywords		
Materials technology, sustainable construction, crop-based materials, building technology, innovation		
Reader interest	**Classification**	
Policy-makers, architects, designers, specifiers, engineers, construction clients, builders, project managers, and cost consultants.	AVAILABILITY	Unrestricted
	CONTENT	Technical guidance
	STATUS	Committee-guided
	USER	Policy-makers, architects, designers, specifiers, engineers, construction clients, builders, project managers, and cost consultants

CIRIA

CIRIA C614   © CIRIA 2004   ISBN: 0-86017-614-2   RP68

Published by CIRIA, Classic House, 174-180 Old Street, London, EC1V 9BP, UK

British Library Cataloguing in Publication Data
A catalogue record is available for this book from the British Library

Acknowledgements

This handbook was produced as a result of CIRIA Research Project 680, "The use of agricultural crops (plant and animal) in construction", by Dr Andrew Cripps of Buro Happold, in partnership with Mr Richard Handyside, Mr Liam Dewar, Mr Jonathan Fovargue of Construction Resources and with assistance from Ms Catrin Sneade, Ms Kathleen Ter and others of Buro Happold. R.J. Rickson and M.J. Hann of National Soil Resources Institute, Cranfield University contributed to the chapter on geotextiles.

Dr Andrew Cripps is a physicist turned engineer who coordinates research, development and innovation across disciplines for Buro Happold. He has a particular interest in environmental issues. Relevant research includes the cardboard school project, reuse of building components and a CIRIA guide to polymer composites.

Mr Richard Handyside established Construction Resources, Britain's first ecological building centre and builders merchant, in the late 1990s. The centre was officially opened in May 1998 and promotes and distributes a wide range of ecological building materials and systems.

Mr Liam Dewar is an architect with experience in both UK and Continental practices, including Herzog De Meuron, and employed until April 2004 at Construction Resources. He now works for Eurban Ltd.

Mr Jonathan Fovargue has worked as a salesman for a variety of manufacturers of construction products. He was sales manager at Construction Resources and now works for Eurban Ltd.

The project was initiated by Mrs Ann Alderson, initially managed by Ms Arna Peric-Matthews and Miss Sarah Reid and subsequently led by Dr Das Mootanah, project manager, CIRIA.

Following CIRIA's policy of collaboration, the study was guided by a steering group of experts involved, or with an interest, in crop-based materials. CIRIA would like to express its thanks and appreciation to all members of the project steering group for their helpful and useful comments and advice. The steering group comprised:

Members

Dr Phil Bamforth (Chair)	Independent consultant
Mr Henry Aykroyd	Biofibres Ltd
Dr David Carmichael	representing National Farmers Union
Mr Ralph Carpenter	Modece Architects
Mr George Henderson	WS Atkins (representing Department of Trade and Industry)
Mr Neil May	Natural Building Technologies
Mr Aarun Naik	Forum for the Future
Mr Andrew Ormerod	Eden Project
Dr Clare Perkins	Arup
Mr Nick Starkey	Government-Industry Forum on Non-food uses of crops
Dr Bruno Viegas	Department for Environment, Food and Rural Affairs
Prof Tom Woolley	Professor of Architecture at Queen's University of Belfast. (Member of the Association for Environment-Conscious Building).

Corresponding members

Mr Peter Allen	University of East London/Centre for Alternative Technology
Dr Mark Hughes	The BioComposites Centre, University of Wales
Dr Derek Stewart	The Scottish Crop Research Institute
Mr Robert West	Qinetiq

CIRIA's project manager was Dr Das Mootanah.

Funders

The project was funded by:

Department of Trade and Industry, through its Partners In Innovation (PII) programme,

Department for Environment, Food and Rural Affairs,

and CIRIA's Core Members.

Thanks to the following who provided photographs/illustrations for this publication:

Arup	Figures 10.8, 10.9
Alistair Bruce	Figure 8.2
Peter Brugge	Figure 8.4
Brunel University	Figures 10.4–10.7
Buro Happold	Figures 2.1, 4.3–4.6, 8.3
Ting Chen	Figure 4.1
Construction Resources	Figures 3.5–3.11, 4.7–4.9, 5.1, 5.4–5.7, 6.3, 6.6, 6.7
Andrew Cripps	Figures 4.10, 5.2, 5.3, 6.1, 6.2, 6.5, 7.5, 7.6, 10.2, 10.3
Gaia Architects	Figures 9.11–9.14, 9.16
Mike Hann	Figure 7.7
Inthatch	Figure 8.1
Maccaferri	Figures 7.3, 7.4
R J Rickson	Figures 7.8, 7.9
Second Nature UK Ltd	Figure 9.15
Mark Stewart Architects	Figures 9.1–9.3
Suffolk Housing Association	Figures 4.11, 9.4–9.10
University of West of England/White Design	Figures 4.2

Executive summary

This handbook aims to encourage the use of products from agricultural crops by those in the mainstream of the UK construction industry. The term "crop" is associated mainly with plants, but in this handbook it is also used to include animal-based outputs from farming, eg wool. The target audience spans the construction supply chain including those who commission, design and build construction projects, and have an interest in reducing the environmental impact of the process.

There is great potential for crops and by-products of animal husbandry to help reduce the environmental impact of construction. At the same time this can also help improve the economic viability of aspects of agriculture in the UK and world-wide, through the possibility of adding value to existing crops, using waste materials or developing new crops.

The main focus of this handbook is on products that are currently available, and how these can be used successfully in construction projects. There is also coverage of products that are being developed and should be available in the near future.

In addition to environmental benefits in production, many of these products bring environmental benefit in use. These relate to better air quality, natural management of moisture levels and reductions in allergic reactions. The "natural building" sector has been aware of these benefits for many years, but only now is it beginning to enter mainstream construction thinking.

Scope

For the purpose of this handbook, "crops" covers all materials deliberately planted or reared on farms, and these need not be only in the UK. However, because there is a substantial amount of literature on timber and related products, these are not covered in this handbook.

Further, in this handbook, construction is taken as covering all contributions to built works, including "civil engineering infrastructure" projects as well as building. The dividing line in buildings has been drawn with floor and wall coverings included, but furniture excluded.

CD-Rom

The CD-Rom (attached to the inside bacl cover of this handbook) contains:

- a promotional PowerPoint presentation that can be used (or adapted for use) at seminars, conferences and exhibitions to introduce and promote the concept of using crops in construction, with examples in use

- a promotional document summarising the use of crops in construction and the benefits to all those concerned with construction and its environmental impact.

Contents

List of tables

List of figures

GLOSSARY AND ABBREVIATIONS

Agro forestry   Cultivated mixtures of trees, crops and/or livestock

Bast fibre   The fibrous outer fraction of straw

Bio-composite   Composite material that will bio-degrade under appropriate conditions

Biodiversity   The term biodiversity was coined in 1985, (abbreviating "biological diversity") and has been defined in many different ways, including, "Total sum of life's variety on Earth, expressed at the genetic, species and ecosystem level"

Biological control   The control of pests, diseases and weeds through the use of other organisms, often natural predators, parasites and diseases

Biomass   The total weight of living material, of all forms

Bio-polymer   A polymer (long chain molecule) made from a plant-based source

Borates   A salt in which the anion contains both boron and oxygen, BO_3, found in preservatives

Butadiene   A gaseous unsaturated hydrocarbon, used in the manufacture of synthetic rubber

Composite   Material made by combining two or more materials, often used as shorthand for fibre reinforced polymer (FRP) composite

Control of Pesticides Regulations (COPR)   The controls on pesticides, set out in the Food and Environmental Protection Act, are detailed and implemented through The Control of Pesticides Regulations 1986 (as amended)

Conductivity (thermal)   The ability which a specified material has to transmit heat from a region of a higher temperature to a region of lower temperature

Cellulose   An insoluble substance which is a glucose polymer and is the main constituent of plant cell walls and of vegetable fibres such as cotton, paint or lacquer, consisting primarily of cellulose acetate or nitrate in solution

Casein   The main protein found in milk and (in coagulated form) cheese

Creosote   A dark brown oil containing various phenols and other compounds distilled from coal, tar, used as a wood preservative – a colourless, pungent, oily liquid distilled from wood tar. Creosote should no longer be used, but may be found on old materials

Crops   Any product from a farm that has been deliberately cultivated. In this handbook it is the general term used to cover plant-based crops and animal-based products, eg wool

Defra   The Department for the Environment, Food and Rural Affairs

Dynamic stiffness   A parameter used to describe the ability of a resilient material to transmit vibration

Dammar   A resin obtained from various mainly Indo-Malaysian trees, used to make varnish

Distemper   A type of paint having a base of glue instead of oil, used on walls

Eco-composite   Form of composite designed for low environmental impact (see also bio-composite)

Embodied energy, primary   The total energy consumed in the production of a material. The official unit for energy is the Joule (J), but the more commonly-known unit, kilo Watt hour (kWh) is used in this handbook, where 1 kWh=3 600 000 J. It is important to note that the delivery of a kWh of electricity will, in general, use more than a kWh of fuel, when compating energy uses.

Emulsifier   A substance that stabilises an emulsion paint

Emulsion paint	A water-based paint which not shiny when dry
Ester	An organic compound made by replacing the hydrogen of an acid by an alkyl or other organic group
FAO	Food and Agriculture Organisation
Fungicide	A pesticide used for controlling fungal growth
Formaldehyde	A colourless and odourless gas made by oxidising methanol, and used in solution as a preservative for biological specimens, and present in some timber-based products, gigatette smoke and some soft furnishings
Habitat	The area of an environment where an organisms lives, feeds and breeds
Herbicide	A pesticide used to control unwanted vegetation (weed-killer)
Hygroscopic	A material which tends to absorb moisture from the air
Ilmenite	Mineral that is the basis of the titanium dioxide used in virtually all white paints, whether natural or synthetic
Insecticide	A pesticide used to control unwanted insects
Intensive farming	A system of farming with the aim of producing the maximum number of crops in a year, with a high yield from the land available, and to maintain a high stocking rate of livestock
Impact sound	– (or structure-borne) sound is sound energy resulting from direct impact on a building element
Interstitial condensation	Condensation occurring within the building fabric, caused by the moisture diffusing through the fabric and being taken below its dew point
Impact sound pressure level	The measurement of sound pressure levels in a receiving room when the floor/ceiling assembly under test, including the presence of a floor covering if applicable, is excited by a standardised tapping machine in the room above
Jute	A rough fibre made from the stems of a tropical plant
Kenaf	A brown plant fibre similar to jute used to make ropes and coarse cloth
Light earth	A mixture of soil (often clay) with straw or a similar fibre, used to produce a building material with lower weight and better thermal performance than the soil on its own
MAFF	The former Ministry of Agriculture Fisheries and Food, and a forerunner to Defra
Normalised impact sound pressure level	The corrected measurement of impact sound pressure level, to take account of the sound absorption of the receiving room
Organic farming	Farming without the use of manufactured chemicals or genetically modified organisms, and conforming to standards enforced by the Soil Association
Pesticide	A pesticide is any substance, preparation or organism prepared or used for destroying any pest
pH	Measure of chemical acidity (<pH7) or alkalinity (>pH7)
Polyamide, polypropylene, polyacryl	Synthetic fibres based on non-renewable oil

Permethrin	A synthetic insecticide of the pyrethroid gas, used chiefly against disease-carrying insects
Recycling	Taking the waste material from some product or process and reprocessing it into the same or another product
Residue	Any pesticide found in a sample, including any specified derivatives such as degradation and conversion products, metabolites and impurities, which are considered to be of toxicological significance. Pesticide residues are the small amount of pesticides that can remain in the crop after harvesting or storage and make their way into the food chain
Resistance, thermal	Building materials present resistance to the flow of heat. The resistivity of the material is inverse to the conductivity
Reuse	Taking a material that has been used, and using it again without reprocessing. It may involve repair or re-shaping
Sd [m] Diffusion-equivalent air-layer thickness	The thickness of a layer of air which has the same diffusion resistance as the given layer of material
Shive	The inner woody core of the stem of a plant (also called hurd)
Sustainability	From "sustain" meaning to hold up, to bear, to support, to provide for, to maintain, to sanction, to keep going, to keep up, to prolong, to support the life of. (Chambers Concise Dictionary)
Sustainable agriculture	The application of husbandry experience and scientific knowledge of natural processes to create integrated, resource conserving farming systems, based on respect for the people and animals involved, which reduce environmental degradation and which promote agricultural productivity and economic viability in both the short- and long-term
Specific heat capacity	Heat required to raise unit mass of substance by one degree of temperature
Sound pressure	The sound pressure due to a source is the amplitude of the pressure vibrations in a sound wave
Sound pressure level	The sound pressure level (decibels) is 10 times the logarithm, to base 10, of the square of the ration of a given sound pressure. The reference level is 20μ Pa
SBR	Styrene-butadiene-rubber, based on non-renewable petrochemicals, and contains chemicals such as stabilisers, fire-retardants, vulcanising agents and softeners
Sisal	Fibres extracted from the leaves of a Mexican succulent of the genus *Agave*, with large fleshy leaves, cultivated for the fibre it yields
Thermal damping	Reducing the temperature fluctuations within a space by evening out the peaks and troughs
Terpene	Any of a large group of volatile unsaturated hydrocarbons with cyclical modules, found in the essential oils of conifers and other plants
U-value	The conductivity of a material; the resistance per unit of depth
VOC	Volatile organic compound – the term used to describe the mixture of organic (carbon-based) gases that are given off by some materials
Vulcanise	Harden, (rubber, or rubber-based material) by treating it with sulphur at a high temperature

HOW TO USE THIS HANDBOOK

The purpose of this handbook is to encourage the use of crop-based materials in construction. It therefore covers the following elements:

- why crop-based materials are a good idea in general
- the products that are available
- case studies of the successful use of these products
- products that are likely to become available.

Because construction is an industry that involves many people who all influence decisions, this handbook is expected to be read by a wide range of people. It will clearly be of direct interest to the architects and engineers who design buildings, and to those clients who are aiming for lower environmental impact buildings. It should also help builders, project managers and cost consultants involved in these projects who may be concerned about how to deliver them.

Within the heart of the handbook there are chapters on the main products currently available. Each chapter explains:

- what the product or class of products is for
- why it is worth considering
- the performance requirements for the type of product
- the performance of the crop-based material
- design issues.

It is not possible to give design guidance on such a large range of different product groups within a single handbook, so that the design sections are able to give only indications of key issues to consider. Further guidance will be available from product suppliers and specialists.

The main product groups considered are those currently available in the UK:

- insulation materials
- light wall construction materials
- paints and wall finishes
- floor coverings and finishes
- geotextiles
- thatch.

The future products section covers materials that could have an impact on UK construction in the future, but have little or no current availability:

- polymers from plant sources
- boards
- bio-composites
- structural bamboo
- expanded starches.

1   Introduction

This handbook covers the uses in construction of materials produced from farms. These are mainly plant-based, and the general term "crops" is used to describe them, although some materials are animal-based and should not strictly be called crops. This introduction outlines the main reasons why crop-based materials are of interest in construction.

Timber products were deliberately excluded from the scope of this work. This is not because they are unsuitable, nor that they are not a form of crop, but because they are well covered in other literature, and in terms of current usage dominate other renewable materials.

Further the handbook is addressed mainly to UK construction and, therefore, concentrates on materials available in the UK. It does, however, include some materials produced outside the UK, but that could be (or are) used here. There may be some areas where these categories become blurred.

1.1   JUSTIFICATION 1: ENVIRONMENTAL IMPACT

Products used by construction

A construction product is any component that goes to make a building, from a brick to a light switch. The UK construction products industry is worth £30 billion annually, accounts for 40 per cent of total construction output and nearly 20 per cent of the UK manufacturing base (CPA, 2002).

In mass terms, it has been estimated that the UK construction industry uses more than 300 million tonnes of material each year, or some 5 tonnes per person (Smith, *et al*, 2002). However this headline figure can be misleading, as it is dominated by quarry products, used mainly as "fill" and concrete. The Mass-Balance Report (Smith, *et al*, 2002), gives a wide range of useful information on material flows, summarised in the table below.

Table 1.1   *Primary materials used in construction*

Product group	'000 tonnes	%
Quarry products	125 871	42.5%
Cement, concrete and plaster products	97 992	33.2%
Stone and other non-metallic minerals products	43 631	14.8%
Wood products	9241	3.1%
Bricks and other clay based products	5979	2.0%
Ceramic products	4313	1.5%
Fabricated metal products	3938	1.3%
Finishes coatings adhesives etc	1477	0.5%
Plastic products	1402	0.5%
Glass-based products	1415	0.5%
Cabling wiring and lighting	190	0.1%
	295 449	100%

Source: Mass-Balance Report (CIRIA, 2002) – (1998 data)

These are all very large quantities of material, and represent a substantial use of materials and hence a very large part of UK energy use and environmental impact.

Waste

In 1998 some 20–25 million tonnes of construction and demolition waste a year went to landfill. This represents between 30 and 40 per cent of total waste (DTI, 2002)).

Additionally, in 2001, the Office of the Deputy Prime Minister (ODPM) carried out a survey into the use of construction and demolition wastes. That survey highlighted that total construction and demolition waste for England and Wales was estimated at 93.91 million tonnes, up from an estimated 72.5 million tonnes in 1999. Of this, 48 per cent was recycled and a further 48 per cent was beneficially reused, mainly for layering or topping at landfill sites and backfilling quarries. The remaining 4 per cent was sent to landfill as waste (ODPM, 2001).

A more efficient use of construction materials in all stages of the construction process would reduce the amount of waste generated. This would maximise the opportunities for greater reuse and recycling of construction materials and, perhaps, encourage the industry to trial other ecologically sustainable construction materials. This greater resource efficiency would help minimise the environmental impacts of construction including through lower demand for virgin material and reduced burden on landfill sites.

Energy

Many construction materials, such as concrete or fired bricks and blocks, use large amounts of energy in their production and transport. (This is known as the "embodied energy" of the material). It has been suggested that 66 per cent of total UK energy consumption is accounted for by construction and use of buildings (Woolley *et al*, 2002). Most of this energy is produced by the burning of fossil fuels, which increases the amount of carbon dioxide (CO_2) in the atmosphere, causing the temperature of the earth to rise which is linked to climate change.

Construction materials made from crops generally use much less energy in production than conventional materials, and if they can be grown, manufactured and used within a small geographical area, then the energy used for transportation would be reduced as well.

The widespread use of agricultural crops would greatly reduce the impact of construction material use. In addition, any waste from agricultural crops can normally be disposed of safely and easily, with little or no environmental damage. These factors combine to provide a substantial environmental benefit to the use of crop-based materials, and others will emerge in the more detailed chapters that follow.

1.2   JUSTIFICATION 2: SOCIO-ECONOMIC BENEFIT

Sustainable development is high on the UK Government's agenda. Increasing the use of natural, easily renewable, and comparatively cheap resources produced from land that is no longer used for food cultivation, will contribute significantly to increased sustainability. It could also help the agricultural industry by providing a good return on land for the farmers.

The Construction Statistics Annual (DTI, 2002) found that the UK is a net importer of £2966 million worth of building materials and components each year. Therefore the establishment of a strong UK production industry could help to reduce this.

The construction industry is one of the largest sectors in all countries. In the UK it accounts for around 10 per cent of GDP (Morton, 2002) and produces physical infrastructure and structures such as offices, houses, hospitals, schools, bridges and roads. In 2001 the value of all the new orders from the private sector was £22 357 million. The public sector made up a further £7286 million (Morton, 2002). In addition there is a similar or larger-sized activity in the repair and maintenance of the existing built environment.

The make-up of the firms within the industry creates several complications apart from its shear size. There is a very large number of small firms, and at least 500 000 sole traders – making up part of a total workforce of around 2 million in the UK (Morton, 2002). At the same time there are some very large companies carrying out the largest projects, but these often sub-contract aspects of the work to small firms. Further there is generally a division of responsibility between design and construction and, often, costing. This sometimes makes it difficult to understand how decisions are made. The construction industry has well-established ways of working and tends to be slow to change.

All of these aspects mean that it is by no means easy to achieve change, but the need to reduce environmental impact is recognised, so the effort needed to impact on the industry may pay off.

On the other hand, it is a significant factor that agriculture and farming is a cornerstone of the rural economy and the major user of rural land.

In 2002, Defra published a strategy for the future of agriculture, entitled "The Strategy for sustainable food and farming – facing the future" (Defra, 2002). This describes options the government will pursue to help achieve its vision for agriculture. This provides a vision that sees an agriculture and farming industry that is vibrant, competitive and sustainable, with high environmental and welfare standards, playing its full part in contributing to the economy, to rural communities and in sustaining an attractive countryside.

Previously the government (MAFF, 1999) encouraged new rural industries based on a plan for the sustainable development of an area. The emphasis was on creating new businesses when possible, and there was scope for enhancing and developing the existing businesses in rural areas.

Consequently, this highlights opportunities available for developing the agricultural industry to expand into and provide the construction industry with agricultural crops that could be used in different aspects of construction material. This could include the production and marketing of new or non-mainstream crops or livestock products, to be used for mainstream construction products. At the same time this can also help to improve the economic viability of aspects of agriculture in the UK.

There are additional opportunities arising from changes in legislation, and the UK landfill tax and aggregates tax are particular examples of government actions that can in some circumstances shift the price balance towards natural and recyclable materials.

1.3 JUSTIFICATION 3: CONSTRUCTION PERFORMANCE

In some cases, and in addition to the benefits of the production process, which are mainly environmental, there are also performance benefits from natural products. These are discussed in detail in the chapters that follow each product area.

However, important examples include:

- natural fibre geotextiles that stabilise soil and allow plant growth, and biodegrade over time so that no plastics are left in the soil or need to be removed when new plants are established

- paints based on a range of natural ingredients, including plant oils and colours, that produce little or no indoor air pollutants during application or normal use

- insulation materials made from natural fibres that provide buffering of moisture and heat, to assist with the indoor environment conditions

- board products that are effective in use but allow easy disposal through composting at end of life.

1.4 AGRICULTURAL ECONOMICS: BARRIERS AND ENABLERS

This section discusses the impacts of agricultural economics and possible changes to agriculture that could affect the availability of construction products. One of the drivers behind a move to crop-based materials is the aim to increase income from farming in the UK, and to reduce the trade deficit in the materials sector. For this to occur it is necessary that the agricultural sector is able to respond and benefit from these opportunities. There is clearly a risk that only agriculture in other countries will benefit.

This section should also help with identifying opportunities for agriculture, and where action may be needed to help the establishment of new products or processes. More ideas for future research and development work are discussed in Chapter 11 (Conclusions and further work).

One of the steering group members for this project, Peter Allen, carried out a study looking into the viability of some products for farmers to produce, (Allen 2003). Working with the National Farmers Union, he surveyed more than 200 farmers, and found that there is wide support for the principle of growing crops for non-food purposes. They were, not surprisingly, nervous of investing in an uncertain idea and this is reflected in the current low level of production of specific crops. But there is clearly the potential for take-up by the farming community when the conditions are appropriate.

This study also reinforces the considerable potential for savings in embodied energy by the adaptation of building technologies to the use of crops. This is driven by the alternatives to energy-intensive bricks in particular, but other materials are also important. A further point from this study is that although there is currently a good level of availability of straw, the market would tighten if there was rapid take up of the materials. For sheep's wool, the supply is driven by the price of meat so, while the market could stand (and would welcome) an increase in the use of wool insulation, it may not be expandable beyond a certain level.

New or unfamiliar products will meet barriers to their adoption in any industry, and the construction industry is certainly no exception. In a highly cost-driven industry, cost effectiveness is critical. Architects and specifiers may be uncertain about performance, liability, and conformity with building and planning regulations. Even if clients and architects are convinced, contractors faced with unfamiliar products, perhaps from unfamiliar suppliers, with questions over reliability of delivery, will frequently add a hefty "uncertainty" premium to their bids, or will simply bring heavy pressure on architects to allow substitution by familiar, conventional products. The construction programme may be extended because of the unfamiliarity of the construction materials.

There are potential barriers in the area of production and manufacture too. New manufacturing processes and plants for new crop-based products require heavy investment, and the availability or absence of government support can be critical. Market prices for agricultural products and the prevailing regulatory and subsidy regime can otherwise make promising materials economically unviable. For example, thermal insulation material is perhaps the most obvious product to manufacture from agricultural crops, with some significant performance advantages over petrochemical or mineral-based materials. But thermal insulation material is, by its very nature, a low-value product, used in bulk, and it may not be possible for European farmers to produce appropriate crops at a price that would make the product commercially viable.

However, where government chooses to take action, it is possible for change to take place. For example Germany is currently providing incentives for the use of crop-based insulation materials.

The following issues arise from the agricultural nature of crops:

- the variability of yields dependent on weather, soil and other environmental conditions
- the availability of crops at limited times of year, resulting in storage requirements
- the low density of most crop material, requiring transport to a processing plant
- uncertainty over future costs dependent on subsidy levels.

Because any new product will inevitably start with a relatively low market share, there is a problem with processing plant availability, even where existing equipment can be used. A good example of this is that MDF (medium density fibreboard), and other boards could be made with a range of fibres, but the manufacturers will not interrupt their main processing to produce a short run of an alternative material. In this particular application the availability of low cost wood waste also prevents the use of other fibres, as these cannot compete on price even though they are technically suitable.

Possibility of local processing

One idea that might help to alleviate the transport problems would be the creation of mobile processing systems. These might allow the processing of boards or other products on or very near to the site of the harvesting, reducing transport costs and double handling. Further research would be needed to investigate the practicality and economic implications of this idea.

Marketing

One clear deficiency at present is any collective marketing of crop-based materials, meaning that users and specifiers will not be aware of the options that are available. As we have seen with organic food, and to some extent the British food "tractor" mark, there is a long-term benefit in marketing a range of products as bringing a collective benefit. This handbook plays a part in this, but it could be taken further in marketing terms.

Genetic modification

It is sometimes suggested that modifying crops to increase yields or modify products would be a benefit, and would accelerate the uptake of crop-based construction products. While it is clearly theoretically possible that this could be the case, it is unlikely to help the introduction of new products. This is because there is extensive consumer resistance to GM crops. This is felt to apply strongly, particularly by environmentally-minded clients and professionals, who are precisely the key target market for products that are meant to be environmentally beneficial. It is, therefore, a more secure approach to avoid GM crops entirely, and consider the additional benefits of organic production of the raw materials.

2 Background to use of crops

There is already a substantial amount of literature on issues relating to crops and their use as construction products. This forms the basis on which new work should build.

2.1 OVERVIEW OF INDUSTRIAL CROPS

IENICA (Interactive European Network for Industrial Crops and their Applications) categorise industrial crops as falling into four main areas (IENICA, 1999):

1 **Chemicals:** polymers and plastics, dyes, paints and pigments, pharmaceutical.

2 **Speciality chemicals:** adhesives, agrochemicals, personal care products, soaps and detergents, specialised organics.

3 **Industrial fibres:** paper and board, composites, textile fibres, bulk fibres.

4 **Industrial oils:** two-cycle oils, transmission fluids, lubricants.

Applications of industrial crops are driven by key factors such as: materials performance, their sustainability, "cost in use", politics, consumerism, "the Green movement" and climate change.

The global market for non-food crops is valued at £20 billion. It is estimated that approximately 1400 organisations are involved or interested in the application and development of alternative crops in the UK, 50 per cent of which are large global companies such as Cargill and British Sugar (IENICA, 1999). Although the number of organisations involved in this area is not expected to increase dramatically it is anticipated that there will be additional requirements for primary processing, information and data, contacts and reference points.

In 2003 the National Centre for Non-Food Crops, NNFCC <www.nnfcc.co.uk>, was launched. It is based near York with the purpose of "providing a single, independent and authoritative source of information on the use and implementation of non-food crop products and technologies in the United Kingdom".

In 1999, in response to a House of Lords Select Committee Inquiry on non-food crops, the Royal Society pointed out the following main opportunities for industrial crops (Royal Society 1999):

- there is significant potential for non-food crops to make an impact on economic activity and land use in the UK

- areas of particular potential are crops intended for energy production or sources of fibres for bio-composites

- crops used for speciality chemicals are likely to become increasingly viable depending on processing technology

- although genetic modification of crops would increase the range of applications, care needs to be given to environmental concerns

- incentives such as subsidies or taxation are limited. Research funding and the coordination of activities are needed to take advantage of the development potential of non-food crops.

HISTORICAL USE OF ANIMAL AND VEGETABLE CROPS IN BUILDING

Some agricultural crops are already used in construction products. One of the oldest and best established uses in the UK is in thatch roofs. Other countries, particularly third world countries, have used natural plant materials extensively in construction for many years.

Additionally, because of its versatility and widespread availability, earth (with fibre) has been used as a construction material on every continent and in every age. Earth construction takes many forms, including adobe, sod, rammed earth, straw-clay, and wattle-and-daub. "Cob" is the English term for mud building, which uses no forms, no bricks, and no wooden structures. Exactly when and how cob building first arose in England remains uncertain, but it is known that cob houses were being built by the 13th century. However it happened, cob houses became the norm in many parts of Britain by the 15th century, and stayed that way until industrialisation and cheap transportation made brick popular in the late 1800s.

Sheep's wool has been used as an insulation for centuries. An example of wool in construction was recently demonstrated on the BBC Restoration programme, which showed Banchory Sanatorium in the Scottish Highlands, a Bavarian-style, timber-framed building dating back to 1900. The sanatorium was designed by George Coutts of Aberdeen and wool was found to be used as the insulation.

Similarly, hemp is known to be the oldest cultivated fibre plant in the world. Hemp fibres have long been an industrial fibre source because of its steady availability, strength and versatility. Materials made from hemp fibre have been discovered in tombs dating back to 8000 BC.

There is a strong use of crop-based materials in what is known as the "natural building" sector, dominated for the main part by self-builders, usually in rural areas and often inspired by the need to produce low cost dwellings. The book *The Art of Natural Building* (Kennedy *et al*, 2002) is typical of the literature that has emerged from this movement. It deals with a wide range of building materials, covering those that are crop-based as well as earth and timber, and features a range of building examples.

The challenge that this handbook is aiming to address is to move the technologies discussed above more towards the mainstream of the construction sector. The cutting edge of green building is vital to keep testing and developing new ideas, but it is the mainstream that needs to change for the largest impacts to occur.

OTHER INDUSTRIES

The automotive industry experimented with the use of natural fibre composites (or bio-composites) as long ago as 1941 when Henry Ford manufactured a prototype car with bodywork manufactured from flax, hemp and other organic materials (Hendley 2001). Today there is a revival in their application within the industry, with the Mercedes E class using jute-based door panels in 1994. Interest focuses primarily on the use of fibres, namely hemp and flax which are grown in western Europe and the sub-tropical fibres, jute and kenaf, which are from the Indian sub-continent.

Reasons for the steady growth in the natural fibres market include:

● comparative weight reduction of 10–30 per cent in comparable parts

● good mechanical and manufacturing properties

- scope for forming complex components in a single machine pass

- relatively good impact performance, with high stability and minimal splintering

- occupational health advantages in assembly and handling compared to glass fibre, where respiratory problems can be caused by airborne glass particles

- moulding off-cuts can be reused, unlike fibreglass

- reduced emissions of toxic fumes when subject to heat

- good green credentials as a sustainable renewable raw material resource

- superior environmental balance during material and energetic use

- recycling possibilities by incineration with energy recovery or by regrinding

- relative cost advantages compared with conventional constructions.

Many of the benefits of using these natural fibres in the automotive industry are equally applicable to the built environment. It is also important to note that not all of the automotive applications have continued to be successful, in part because the composite panels cannot be recycled, and this is an important requirement for the End of Life Vehicle Directive. Constraining factors include:

- concerns regarding the quality of bast fibre supply and its long term availability

- technical problems relating to variation in quality of different batches, or emission problems

- difficulty of recycling of mixed materials, and the impact of this on the ability to meet the End of Life Vehicle Directive.

2.4 THE MARKET

Market information for crop and animal products is hard to obtain, reflecting the modest development of the sector. Some information is given in the main chapters that follow but in general, the market share of these materials is very small.

It is clear that some other countries, particularly Germany, have already made greater use of crop-based materials in construction. There is much to learn from their experience, although most of their literature is not available in English. Many of the products that are available in the UK are imported from continental Europe, and this added cost (due to the transport of material) inevitably reduces the competitiveness of the products.

This handbook focuses on the applications of materials by looking at the product groups, because this reflects the ways in which the design and construction process can best use the information. However, it can also be useful to consider applications categorised according to source material as shown in Figure 2.1. This expresses the range of materials currently being used or developed, according to their agricultural source. All the classes of crop are being used to some extent, but there will be opportunities for development.

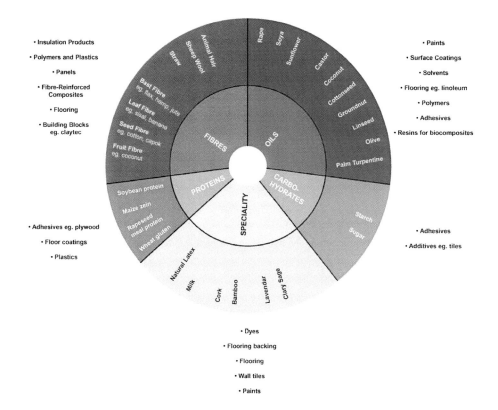

Figure 2.1 *Applications of crop material by source*

3 Insulation

This chapter presents information on crop-based insulation materials, and where and how they can be used successfully in buildings.

The table below summarises the contents of the chapter. The following sections explain the nature of insulation materials, the different roles they have to play and their appropriateness for each location.

Table 3.1 *Properties needed for different uses of insulation*

What/where	Properties needed	Crop-based options	Other natural options
Horizontal insulation (loft), for thermal properties	Maximum heat insulation, minimum cost, (preferably lightweight)	Hemp, flax or other fibres Wool	Recycled paper
Vertical (wall) or sloping for thermal properties	Thermal insulation, self supporting rigidity	Fibre batts (wool, hemp, flax etc). But needs a breathing wall construction	Recycled paper
Horizontal under-floor insulation (load-bearing)	Thermal insulation, compression strength	Probably none – not robust enough	
Cavity fill insulation	Thermal insulation, able to be injected	Probably none	
Acoustic insulation – flow resistance	Density, low stiffness	Hemp/flax batts, wool	
Acoustic insulation – dynamic of impact resistance	Density, stiffness optimised – may carry some load	Coconut fibre, hemp/flax boards, straw boards, wool	
Fire protection	Resistance to fire, resistance to flame spread	Cork, reed, compressed straw, wool. Others with treatment	

In practice the thermal insulation will provide some acoustic insulation, and vice versa.

3.1 WHAT ARE INSULATION MATERIALS FOR?

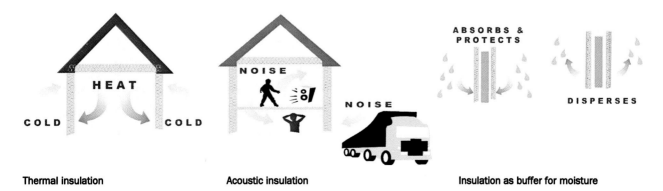

Thermal insulation   Acoustic insulation   Insulation as buffer for moisture

Figure 3.1 *Purposes of insulation*

An insulation material is a material whose primary purpose is to prevent unwanted movement of heat or sound. As illustrated by Figure 3.1 above, insulation materials are included in buildings to carry out one or more of the following roles:

- keeping heat inside the building in cold weather

- keeping the inside cool in hot weather

- limiting incoming external noise

- limiting internal noise transmission.

In addition insulation can also be needed to:

- absorb or re-emit (buffer) moisture from the air to protect other materials

- store thermal energy

- carry limited structural loads (particularly under floors)

- protect against fire spread.

Since energy consumption was not widely considered to be an important issue until the early 1970s, the manufacture and use of materials specifically designed to thermally insulate buildings is relatively new. Insulation materials are limited to use in buildings rather than construction in general. To distinguish crop-based insulation materials from other insulation products it is helpful to group insulation materials according to their material composition as follows:

- mineral fibre insulation (stone wool, glass wool)

- other mineral insulation (ie expanded clay, perlite, cellular glass, etc)

- polymer foam insulation (polyurethanes, polystyrene, etc)

- plant-based insulation (wood fibre, recycled cellulose, flax, hemp, cork, etc)

- animal-based insulation (sheep's wool).

3.2   WHY CONSIDER NATURAL FIBRE INSULATION?

There are three main aspects to the case for and against crop-based and natural fibre insulation:

- in their production they have low environmental impact

- in use they can perform better than competing materials

- in first cost terms they are generally more expensive.

These features have defined the current market for the products. This results in a relatively small niche of consumers with a particular concern about environmental issues, or particular technical needs in the areas of moisture or acoustics. These clients judge the materials to give good value for the higher initial cost being paid.

Each of these features is developed in more detail throughout this chapter.

3.3 UK MARKET

Mineral-based insulation products (stone wool, glass wool) represent about two thirds of the total volume of insulation used in UK construction. Polymer foam insulation covers most of the remaining third of the market.

Even in countries such as Germany and Austria, where the uptake of renewable insulation products has been encouraged, they represent only 3–5 per cent of their respective insulation markets, most of which is timber-based, that is recycled newsprint, wood fibre and wood-wool boards.

3.4 PERFORMANCE CHARACTERISTICS OF MATERIALS AND PRODUCTS

Note that throughout this section all data are taken from manufacturers' information, as there is only very limited formal published information in this area.

3.4.1 Thermal performance – heat transfer
Building Regulations Part L

Thermal insulation materials are materials that have poor heat transfer properties. A material can be considered a thermal insulation material if its thermal conductivity is less or equal to 0.1 W/mK.

The following data presented in Table 3.2 are for "dry" insulation measured according to BS EN ISO 8990:1996, Thermal insulation – determination of steady-state thermal transmission properties – calibrated and guarded hot box. A discussion of the impact of moisture on insulation follows, because it is an important difference between the materials.

Material	Conductivity W/mK	U value of 150 mmW/m²K	Density Kg/m³
Sheep's wool	0.037–0.039	0.25–0.26	16–25
Flax	0.037–0.042	0.25–0.28	20–30
Hemp	0.038	0.26	15–60
Rock wool (values from CIBSE, 1999)	0.033 – 0.047	0.23-0.31	23–200

Table 3.2 *Typical thermal performance data for different insulation materials*

The key terms here are conductivity, resistance and U value, defined in the Glossary. The thickness of insulation required will be dependent on the conductivity. The U values must be compliant with Part L – Conservation of fuel and power of the Building Regulations, which ensures new buildings are reasonably thermally efficient, and these vary with building type.

When calculating for open blown loft spaces or partial-fill floor cavities, it is also important to use the settled thickness (typically 0.8 or 80 per cent of the installed thickness) in the calculation.

In designing, it is important to use the material efficiently, considering future possible reuse. This is not as important as with conventional insulation as the embodied energy is lower.

Thermal insulation: effect of moisture

Most insulation works by trapping air inside the framework of the material. Air has a very low rate of conduction of heat, and by holding the air, convection of heat is largely minimised. However water is a good conductor of heat so, if the insulation material becomes wet, heat will pass through it more easily, and the claimed figures for insulation may not be realised.

There is, surprisingly, little information available on the impact of moisture on the thermal performance of insulation materials, and this is an area that requires further work. A good example of the impact is found in the Haverhill Houses reported in Chapter 9: (Case studies). Although it cannot be proved that it is because of moisture, the work found that although the predicted energy use was lower for the mineral wool houses, the actual use was lower in the hemp houses. The predictions are based on dry hot box tests of insulation, and perfect installation, and the reality of damp insulation is likely to be the cause of the difference.

Figure 3.2 (Sarmala, 1984) below, shows the change in conductivity in an insulation material as the moisture level increases. It is clear that increasing moisture content reduces the resistance of the insulation material to the movement of heat. This effect is seen to be greater at higher temperatures, presumably because the moisture is freer to move faster when it is warm, and therefore transfers heat faster. There is a lack of knowledge of the moisture levels that do occur in the insulation of real buildings, and hence of the extent this effect could be having on energy use.

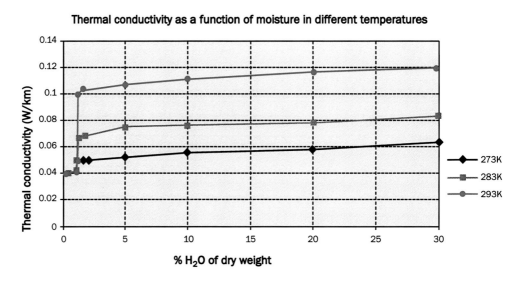

Figure 3.2 *Thermal conductivity as a function of moisture and temperatures.*

The use of in-use figures for conductivity is not uncommon in other European countries. For example, test figures in countries such as German, Austria and Switzerland are increased by a certain percentage to take into account the reduction in thermal performance of an insulation material as a result of moisture content and durability.

3.4.2 Thermal performance – heat storage capacity

The specific heat capacity and the density of a material determine its ability to store heat. The specific heat capacity multiplied by the density, gives the heat storage capacity that is then used to calculate thermal damping. The more heat a material can store the slower it will respond to heating or cooling, and the easier it is to create stable room environments. Mineral-based materials have specific heat capacities of around 1.0kJ/kg.K. Plant-derived materials have specific heat capacities of around 2.0kJ/kg.K. This means that a plant-derived insulation material will store twice as much energy as a mineral wool insulation product of equivalent density and thickness. This is particularly significant in roof spaces and prefabricated timber constructions where the insulation material is often the only material that is able to lower the temperature swings on the inside face of the building element and in so doing, thermally stabilises the internal environment.

3.4.3 Acoustic performance – sound transfer
Building Regulations Part E

Acoustic insulation materials are materials which, due to springiness and acoustic absorption properties, increase acoustic insulation (mostly in association with other materials) and/or absorb or dampen the sound in rooms or cavities. Generally speaking, the softer, the more flexible and the denser the insulation material, the better its acoustic performance. In contrast, rigid insulation products can worsen acoustic performance.

Acoustic performance is required between outside and inside, ie to protect the internal environment from external noise such as traffic. Acoustic performance is also required between individual dwelling units, between rooms within a dwelling unit, and within the room itself. Poor acoustic performance is generally considered to be a health risk, as well as a significant irritant.

There are three main types of acoustic performance to consider: airborne sound, impact (structure-borne) sound, and room acoustics. Room acoustics are not directly affected by insulation, as they are defined by the surfaces and finishes of the room, but the other two aspects are discussed below.

Airborne sound – flow resistance

Airborne sound is sound energy that is transmitted through air. Airborne sound reduction through a wall or floor is the quantity of sound absorbed by that element, and the greater the insulation, the better the performance. A good indication of a product's ability to absorb sound energy is its specific flow resistance. The larger the flow resistance, the better the sound absorption. Insulation materials with sufficient specific flow resistance will improve the acoustic performance of a stud wall by 10–15 dB.

Flexible fibrous insulation materials tend to have good flow resistances, and are therefore ideal for use within constructions to reduce airborne sound transfer. Flexible fibrous materials that are also dense, exhibit better performance at lower frequencies than their less dense counterparts. The denser and thicker the material, the better its performance. Materials suitable for use in partitions have a specific flow resistance greater than 5 KNs/m^4. Suitable crop-based materials include insulation batts made from flax, hemp, and sheep's wool. Figure 3.3 below shows that at low frequencies the denser flax fibre is better able to absorb sound than the mineral wool. This is reversed at higher frequencies, but these are of lower importance, as in both cases the absorption is high, and low frequency sound is more often a problem.

Comparison of acoustic absorption of mineral wool and flax fibre insulation

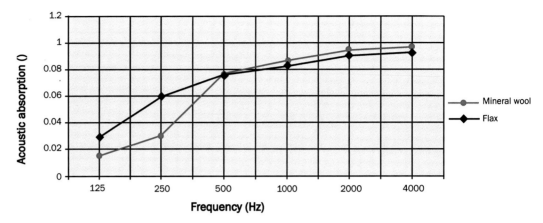

Figure 3.3 *Comparison of the acoustic absorption of a natural and mineral fibre*

Acoustic absorption is the proportion of the sound that is prevented from reaching the other side of a material, so that a figure of 1 is complete absorption.

Impact sound – dynamic stiffness

Impact (or structure-borne) sound is sound energy resulting from direct impact on a building element, and the representative figure is the sound transmission through that element. The smaller the number is, the better the performance. Note that the dB figure for "improvement" given by certain manufacturers is a reduction in impact sound transfer (an improvement on the same floor without the material).

Corrected impact sound pressure level of floor against stiffness of resilient layer

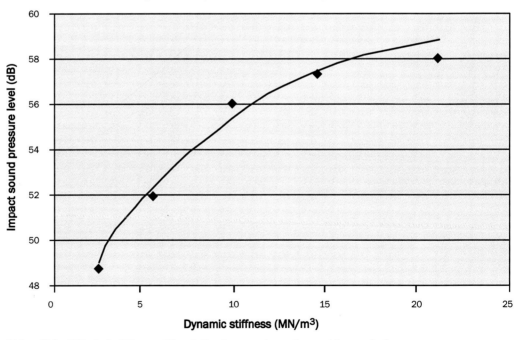

Figure 3.4 *Effect of stiffness of insulation layer on impact sound transmission*

To improve impact sound performance it is necessary to install a resilient layer, a layer that isolates a floating layer such as a screed from a base layer. Insulation materials that are suitable for use as a resilient layer under a floating layer need to have low dynamic stiffness.

Dynamic stiffness is a parameter used to describe the ability of a resilient material to transmit vibration. The variation of sound impact pressure level with the stiffness of the material is shown in figure 3.4. This clearly shows that less stiff materials achieve a greater reduction in transmission. Materials suitable for use as resilient layers have a dynamic stiffness less than 10 MN/m^3. Suitable crop-based materials include coconut fibre, hemp, and flax boards.

3.4.4 Moisture performance
Building Regulations Part C, F

Water vapour diffusion

The water vapour diffusion coefficient of a building material is a measure used to describe a material's resistance to water vapour diffusion relative to that of air. The lower the value, the more permeable a material is to water vapour. Fibrous materials have water vapour diffusion coefficients between 1 and 2, fibrous boards between 3 and 7. Water vapour diffusion decreases with increased moisture content, as there are fewer gaps for moisture to move through when the fibres have swollen.

Sd values (measured in m of air that gives the same diffusion resistance) are a measure of water vapour diffusion for a specific product, in other words, a material with a particular thickness. Vapour impermeable materials (ie vapour barriers) have an Sd-value ≥15 000 m. Insulation materials that act as vapour barriers include cellular glass boards. Material with vapour resistance 2 m < Sd-value < 1500 m are vapour control layers. Insulation materials that act as vapour control layers include most plastic insulation products. Vapour permeable materials (ie breather membranes) have an Sd-value ≤ 0.3 m. Most fibrous insulation materials including crop-based materials are vapour permeable.

When installing crop-based insulation materials between structural timbers, the moisture content of the timbers should be less than 20 per cent before the internal lining is fixed in place. When installing crop-based insulation materials between rafters in pitched roof constructions, the external lining should have an Sd-value ≤ 0.1 m while the internal lining should have an Sd-value ≥ 2.0 m. This is to allow moisture to leave the outer surface of the insulation, but to restrict moisture from the inside reaching it.

Moisture storage capacity

A typical crop-based insulation material is able to store between 20 and 30 per cent of its own weight in moisture, while animal-based insulation material could reach 40 per cent. The insulation material acts as a "reservoir". For denser products, this represents a moisture storage capacity of 30 litres of water per cubic metre. A synthetic mineral wool insulation product on the other hand is able only to store 2.5 per cent, equivalent to 0.5 litre of water. To prevent water damage within a construction, insulation should be capable of storing both the 0.5 kg moisture per m^2 entering a construction during the summer, and the moisture evaporating from timbers which have become wet during installation or have been installed with moisture contents greater than 20 per cent.

Because of this capacity to absorb moisture, natural fibres can help protect sensitive materials from moisture. These might include some timbers and old or valuable materials, eg paintings or plaster.

Interstitial condensation

Interstitial condensation is a build up of condensation within a construction. It is important to ensure that there is not a net gain of condensate over the year. In addition, the condensate shall not adversely affect the long-term durability of the construction, nor shall it adversely affect the performance of the building envelope (increase in thermal conductivity of insulation materials).

Crop-based (and other natural fibre) insulation materials can absorb useful quantities of moisture without it greatly affecting their thermal performance. This property of natural fibre insulation materials is valuable in reducing the likelihood of interstitial condensation. Moisture can be absorbed and then diffuses away to a dryer zone, as long as a breathing construction is used.

Condensation forming within the insulation layer during winter, and the effect of this condensation on the performance and durability of the insulation material, is critical in terms of external wall insulation. For internal wall insulation systems the critical issue is the increased likelihood of condensation forming on the inside surface of all cold bridging, ie at floor/ceiling level, and the decreased thermal performance of the remaining wall due to its inability to dry out properly.

3.4.5 Structural performance
Building Regulations Part A

Most insulation materials are not required to play any role in the structure of the building. However some structural performance is required for each use of insulation. A useful summary of this is provided in several central European countries through a system of abbreviations for use in their respective standards. Since flexibility/springiness can be considered a lack of compression strength, this system also enables the selection of appropriate acoustic insulation materials. The system in use in Germany contains the following abbreviations:

W, WL	without compressive strength, ie between floor/ceiling joists, rafters etc
WD	with compressive strength, but without acoustic isolation properties ie below ground floor screed, below single ply roofing membrane
WV	with racking strength, ie thermal linings, external wall insulation board to take render finish
WB	with bending strength
T, TK	with compressive strength, with acoustic isolation properties ie resilient layers in floor constructions.

3.4.6 Form stability

Air gaps between insulation and timber structure will compromise the overall performance of the building element being insulated in terms of its acoustic, thermal, and moisture performance (the latter resulting in potential loss of structural integrity). Sufficient elasticity and ensuring against unwanted settlement are, therefore, vitally important to the overall performance of the insulation material and the construction. Typical reasons for having air gaps between insulation and rafters/studwork are a lack of flexibility (material too rigid), settlement (insufficient density), slumping (insufficient rigidity) and material shrinkage (resulting from ageing or moisture).

3.4.7	Fire performance
	Building Regulations Part B

Most insulation materials are not used as surface layers of inhabited rooms, so should not represent a fire risk. In some cases they are required to form part of a fire resistant construction. The mineral wool insulation products perform particularly well in this respect as they are non-combustible. Natural fibre insulation materials should not be used in fire resisting construction unless testing to BS 476 parts 20–22 indicate that they perform as required.

Most non-mineral products contain additives to improve their fire performance, ie sufficient to prevent them from burning with flames. The fire retardants most commonly used in crop materials are borates (flax, wool), phosphates (flax, cork, coconut fibre), and soda (hemp, wool). Certain materials (cork, reed, compressed straw, etc) do not require the addition of fire retarding additives, although this can be very much determined by their intended use. Sheep's wool is reported not to burn but to melt away from an ignition source and extinguish itself.

Most crop-based insulation materials contain inorganic fire retardants which inhibit flaming and smouldering combustion. As with most insulation products, precautions must be taken to protect the insulation from heat generated from flues and recessed light fittings. It may also be necessary to de-rate electrical cables that are completely surrounded by insulation. Guidance should be sought from a qualified electrician.

Fire performance of insulation in cavities should comply with requirements of AD "B".

3.4.8	Environmental information

Composition *(toxicological information on ingredients)*

Although the raw materials of most natural fibres, fruits, and bark suitable for use as insulation materials are non-toxic, some of the additives used within the building material products themselves are highly significant in terms of their toxicity. To understand the material, it is necessary to obtain a full material declaration from the manufacturer or supplier of the product. The provision of this information is already an industry standard in certain product areas (for example, natural paint manufacturers) and married to specification writing in countries such as Switzerland.

Typical additives within crop-based insulation materials are borates, sulphates, phosphates, starches and soda. The additives are significantly less toxic and are likely to have lower embodied energy than those typically used in mineral wool products (ammonias, formaldehydes, ureas, phenols, phenolic resins, etc) and those typically used with plastic insulation products (anilin, chlorides, isocyanides, formaldehydes, benzyls, phosphates, phosgenes etc).

Handling

In general, natural fibre insulation brings a benefit over mineral wool insulation in handling, as the fibres are non-irritating.

Because of their form, the cutting of coconut fibre boards is difficult, and the cutting of expanded cork boards is very time consuming. The dust propagation for both products can be annoying and, particularly with cork, may also be a health risk, so that appropriate personal protective equipment should be used, in a similar way to working with timber.

Disposal considerations

It is a major benefit of natural fibre insulation materials that they bio-degrade at the end of their life, and so could be composted or burned for energy recovery. Additives used to limit fire or insect attack may affect these options and so should be minimised.

Natural fibre insulation products containing polyester fibres to bind them, are difficult to dispose of, and may be best burned or recycled. Because of the polyester they are not sufficiently biodegradable for composting and they contain too much organic matter to be ideal for landfill. Mineral wool insulation materials are classified as hazardous waste in certain European countries, and do not biodegrade.

3.4.9 Sourcing/availability

Natural insulation materials, including those that are crop-based, are available in the UK through a small network of ecological builders' merchants. Most crop-based insulation materials are imported, although an increasing number of organisations/suppliers and manufacturers are investigating the possibility of harvesting/manufacturing products in the UK:

- reed boards are currently imported into the UK from central and eastern Europe
- cork insulation products are imported from the Iberian Peninsula
- coconut fibre boards imported from central Europe
- sheep's wool insulation products are produced in the UK from British hill sheep wool, as well as imported from central Europe
- flax and hemp insulation products are imported from western and central Europe
- straw insulation is mainly in the form of locally-grown straw for use in straw-bale building construction. Neither cork nor coconut fibres are suitable for growth in the UK.

3.5 DESIGN ISSUES

Because there are so many different products and applications, it is not possible to give detailed design guidance in this handbook. Advice should be sought from the materials suppliers, and a competent installer is recommended. The basic principles have been discussed in the preceding sections, but some key issues are included here. The most important point is that the design of the building work needs to be adapted to the materials being used, if their maximum performance is to be achieved.

Section 3.6 gives performance data for a range of insulation materials in the form of data sheets.

3.5.1 Specification

The fact that insulation materials are typically hidden from view once installed, encourages poor workmanship and an indifferent approach to the actual specification of the material itself. This has not been helped by the use of standard architects' specification clauses which marginalise the importance of insulation materials by including them under the general heading of "sundry items".

The type of insulation material being specified or chosen tends to be habitual rather than considered on each occasion. The thickness of the chosen material is usually the minimum required to satisfy the static calculation methods within Approved Document Part L, which, as on-site testing has shown, may fail to provide the expected and statutory necessary performance.

3.5.2 Aspects affecting the price

Many products are currently being imported into the UK, and so material supply prices fluctuate due to:

- crop prices
- exchange rate fluctuations
- contractor charges
- sub-contracting and sub-sub-contracting the works.

The scale, and nature of the project itself will influence the cost, as will regional variations, and current work-loads.

To encourage the use of crops-based insulation materials, some countries (notably Germany) have introduced financial incentives in the form of subsidies for use of approved materials. The subsidy currently being offered in Germany is 40€/m^3 of insulation product.

Installation cost can be derived from the following guide installation times for 100 mm thickness:

- loose fill (blow-in) insulation = 0.15 hours per m^2
- insulation batts (between studs) = 0.2 hours per m^2
- board installation time = 0.25 hours per m^2.

3.5.3 Disposal costs

The cost of end-of-life disposal is becoming an increasingly important consideration. Aggregate tax, for example, has been introduced recently to encourage the use of recycled material, and should benefit natural fibre insulation. In addition the possibility of composting crop-based insulation means that the cost of landfill tax (currently £2 per tonne for inactive waste) could be avoided.

3.5.4 Details

Given the many different possible constructions, it is not appropriate to try to produce appropriate "standard details" in a handbook like this. The principles discussed in the preceding sections need to be thought through enough to ensure an appropriate solution, or advice can be obtained from suppliers or designers. The key issues to think through are where the moisture generated inside the building will go, how it may pass through the insulation, and what may happen to any water that gains access from outside of the building.

3.6   PRODUCT DATA SHEETS

Note: All the data in this section is from suppliers' information.

3.6.1   Coconut fibre products

Areas of use
Acoustic insulation (resilient layer) below screeds, acoustic isolation strips for timber/metal stud partitions and timber flooring battens; sealing strip between building components, ie around doors and windows.

Product details
Material:	Coconut fibres (*Cocos nucifera*)
Additives:	Natural latex, mineral salts
Form:	Semi-rigid boards, supplied in packs on pallets
Thickness:	13, 18, 23, 28 mm
Country of origin:	Sri-Lanka (also India)

Material properties
Product type:	Board	
Use classification:	W, WL, T, TK	(DIN 4108)
Thermal conductivity:	0.045	W/m K
Specific Heat capacity:	–	J/m^2 K
Diffusion resistance factor:	$\mu = 1$–2	(DIN 4108)
Bulk density:	50–140	kg/m^3
Dynamic rigidity:	23	MN/m^3 (18 mm thickness)
Fire classification:	B2	(DIN 4102)

Ecological properties
Advantages: Good acoustic properties, no toxic emissions, neutral smell, moisture resistant, highly rot resistant, good form stability, resistant to insect attack.

Disadvantages: Sulphates (used as additive) can be irritants to eyes and skin. Large transportation distances (not possible to grow crop in UK).

Application
Installation: –

Durability: Coconut fibre is extremely tough and rot resistant. Naturally durable.

Removal: Reusable depending on application.

Disposal: Disposal through incineration.

3.6.2   Cork/recycled cork products

Manufacture

Cork is harvested as the bark of the cork oak grown on plantations. The bark is removed every 7 to 9 years and is used in either a granular form, as a loose fill insulation, or can be manufactured into rigid boards by steam-heating the granulated cork to around 380°C. In this process the 2–6 mm granules are bonded together with their own resin.

Primary embodied energy: 40 kWh/m^3

Figure 3.6 *Cork board*

Areas of use
The principal area of application in the UK is as a thermally insulating underlay, under built-up roofing systems or profiled metal deck roofs. Cork board can also be used as an internal lining board, an external wall insulation board, as a resilient layer below machinery, and as a structural isolator.

Product details
Material:	Natural/recycled cork granulate (*Quercus suber*)
Typical additives:	Cork resin (natural) or isocyanate containing adhesives (recycled)
Form:	Board, supplied in packs on pallets
Thickness:	10–300 mm
Country of origin:	Portugal (also Spain, northern Africa, Sicily)

Material properties
Product type:	Board	
Use classification:	WD	(DIN 4108)
Thermal conductivity:	0.050	W/m K
Specific Heat capacity:	1700–1800	J/m^2 K
Diffusion resistance factor:	$\mu = 5$	(DIN 4108)
Bulk density:	110–140	kg/m^3
Compressive strength:	1.5	kN/m^2
Fire classification:	B2	(DIN 4102)

Ecological properties
Advantages:	Chemically neutral, resistant to rodent attack, resistant to rotting, mould resistant, non-electrostatic, good thermal performance during summer due to high thermal mass.
Disadvantages:	Avoid boards bound together with adhesives containing isocyanate and boards which have been factory-bonded to rigid polyurethane foam as there are disposal issues associated with composite materials.

Application
Installation:	When used as external insulation the boards are usually fixed directly t to walls with a mineral adhesive in combination with mechanical fixings. When installing as flat roof insulation, the boards must be kept dry since wet boards can harbour quantities of mould which may lead to allergic reactions. Dust inhalation should be avoided.
Durability:	Naturally resistant against insect/rodent attack and rots with difficulty, even when wet. Being more chemically inert than most other insulation materials, cork is able to withstand deterioration through age.
Removal:	Adhesive fixed and rendered boards cannot be reused easily but dry cork boards can be recycled as loose fill insulation.
Disposal:	No disposal issues, can be composted.

3.6.3 Expanded rye products

Manufacture

Rye grain is mechanically harvested, mixed with mineral additives and manufactured into the final product in a patented extrusion process.

Primary embodied energy: 20–25 kWh/m^3

Figure 3.7 *Expanded rye*

Areas of use
Thermal products for closed horizontal/inclined cavities ie between timber floor/roof joists. Acoustic products for levelling floors in new buildings, for use in renovating old buildings, and as impact sound insulation.

Product details
Material:	Rye grain (*Secale cereale*)
Typical additives:	Lime, mineral salts
Form:	Loose fill, supplied in bags, big-bags or silos
Grain size:	2–4 mm
Country of origin:	Germany

Material properties
Product type:	Thermal fill	Acoustic fill	
Use classification:	W, WD	T, TK	(DIN 4108)
Settlement:	2.6	5	%
Thermal conductivity:	0.045	0.070	W/m K
Specific Heat capacity:	–	–	J/m^2 K
Diffusion resistance factor:	μ = 1	μ = 1	(DIN 4108)
Bulk density:	105–115	210–230	kg/m^3
Compressive strength:	0.04	0.18	N/mm^2
Fire classification:	B2	B2	(DIN 4102)

Ecological properties
Advantages:	No preservatives, no insecticides, no heavy metals. Minimal impact during manufacture. No toxic emissions.
Disadvantages:	Susceptible to moisture damage. Not suitable for use in areas with high humidity or applications where it may get wet.

Application
Installation:	Product can be poured in or blown in using a blowing machine. Loose fill acoustic fill product is not suitable for use in the floors of bathrooms and kitchens. Allow for settlement in the design process. Wear a dust mask during installation. Protect from water.
Durability:	New material with minimal case history. Durability is unknown. Use in low-risk constructions.
Removal:	No known issues.
Disposal:	Products are reusable, compostable, incinerable.

3.6.4   Flax/flax-rich insulation products

Manufacture

Flax fibre is mechanically harvested, left to rot, pressed into bales, broken, stripped, carded, treated with mineral/mineral salts, formed into batts in a mechanical process. Products may contain up to 20 per cent synthetic support fibres (non-recyclable compound).

Primary embodied energy: 70–90 kWh/m^3

Figure 3.8 *Flax insulation*

Areas of use

Thermal and acoustic insulation for internal walls and floors; thermal insulation between wall/roof timbers; felts as acoustic underlay for timber floors; strips as draft-proofing/sealing strips between building components, ie around windows and doors.

Product details

Material:	Flax fibres (*Linum usitatissimum*)
Typical additives:	Polyester fibres, potato starch, mineral salts
Form:	Board (board, batt), roll (quilt, felt, strips), loose (wadding)
Thickness:	50–100 mm (quilt), 40–160 (batts), 2–10 mm (felts)
Country of origin:	France, Germany, Austria

Material properties

Product type:	Quilt	Batt	Felt	
Use classification:	W	WL, WV	T, TK	(DIN 4108)
Thermal conductivity:	0.040	0.040	0.050	W/m K
Specific Heat capacity:	1550	1550	1550	J/m^2 K (flax)
Diffusion resistance factor:	μ = 1-2	μ = 1-2	μ = 1-2	(DIN 4108)
Bulk density:	20	30 – 40	130 - 160	kg/m^3
Dynamic rigidity:	–	–	–	MN/m^3
Fire classification:	B2	B2	B2	(DIN 4102)

Ecological properties

Advantages:	Made from renewable material, low energy during manufacture, non-irritant to skin, eyes or respiratory system, moisture regulating qualities, reusable. CO_2 neutral.
Disadvantages:	Synthetic support fibres (energy intensive manufacture, oil-derived product). The use of herbicides during growing can have a significant environmental impact.

Application

Installation:	Batts are cut with a serrated knife or handsaw, wedged between the studs, joists. Additional fixing with hand tacker may be required. It is recommended that wadding is treated with 3–5 per cent lime before installation to prevent biodegradation. Protect from water.
Durability:	Insufficient information available. Use in low-risk constructions.
Removal:	No issues. Can be reused.
Disposal:	Products can be returned to manufacturers for recycling. Compostable if mixed with organic materials. Products containing synthetic supporting fibres may require incineration.

3.6.5   Hemp/hemp-rich insulation products

Manufacture

Hemp fibre is harvested, dried in the field, pressed into bales, broken, stripped, and undergo a decortication process where the outer fibre layer is stripped from the inner wooley core, usually referred to as shive. Typical yields are around 6–13 t/ha. Fibre-based products may contain up to 15 per cent synthetic support fibres (non-recycleable compound).

Primary embodied energy: 30 kWh/m^3

Figure 3.9 *The Hemp plant*

Areas of use
Thermal and acoustic insulation for internal walls and floors; thermal insulation between wall/roof timbers; acoustic underlay for timber floors; as sealing strip between building components. Hemp shives are used as insulant in lime/hemp mixtures.

Product details
Material:	Hemp fibres and/or shives (*Cannabis sativa*)
Typical additives:	Polyester fibres, sheep's wool, mineral salts
Form:	Board (board, batt), roll (quilt), loose (wadding, shives)
Thickness:	40–240 mm (batts, quilt), 15–30 mm (boards)
Country of origin:	France, Germany, Austria

Material properties
Product type:	Loose	Batt	Board	
Use classification:	W	WL, WV	T, TK	(DIN 4108)
Thermal conductivity:	0.065	0.045	0.040	W/m K
Specific Heat capacity:	2100	2100	2100	J/m^{2}K (hemp)
Diffusion resistance factor:	μ = 1-2	μ = 1-2	μ = 1-2	(DIN 4108)
Bulk density:	–	15–60	90–130	kg/m^3
Dynamic rigidity:	–	–	8–21	MN/m^3
Fire classification:	–	D-s3/2, d0	E	(EN 13501)

Ecological properties
Advantages:	Because of the vigour of the plants, hemp plantations do not require herbicides and pesticides. Hemp fibres are naturally resistant to fungal and bacterial attack, therefore do not require impregnation. CO_2 neutral.
Disadvantages:	Synthetic support fibres (energy intensive manufacture, oil-derived product).

Application
Installation:	Batts are cut with a serrated knife or handsaw, wedged between the studs, joists. Additional mechanical fixing with hand tacker may be required. Protect from water.
Durability:	Insufficient information available. Use in low-risk constructions.
Removal:	No issues. Untreated hemp can be reused.
Disposal:	Products can be returned to manufacturers for recycling. Products containing synthetic supporting fibres may require incineration.

3.6.6   Reed insulation products

Manufacture

Reed grows predominantly in central Europe on the edges of still waters and rivers. The plants can grow up to a height of 4 m. After being cut, the reed is left to dry. The reed is sorted by hand and then bundled in 1 m diameter bundles. Process: Cut, dried, sorted, bundled, mechanically pressed, bound (with galvanised wire), cut to size.

Primary embodied energy: 5 kWh/m^3.

Figure 3.10 *Reed board*

Areas of use
Used externally and internally as an insulating substrate board for plasters and renders (in particular, clay plasters); also used as permanent shuttering for walls and arched floors. Reed mats are used as an alternative to laths in plasterwork.

Product details
Material:	Untreated reed (*Phragmatis australis*)
Typical additives:	Galvanised wire
Form:	Semi-rigid boards, supplied loose on pallets
Thickness:	20, 50 mm
Country of origin:	Central Europe

Material properties
Product type:	Board	
Use classification:	WL, WV	(DIN 4108)
Thermal conductivity:	0.060	W/m K
Specific Heat capacity:	–	J/m^2 K
Diffusion resistance factor:	μ = 2–5	(DIN 4108)
Bulk density:	190–225	kg/m^3
Fire classification:	B2	(DIN 4102)

Ecological properties
Advantages:	Natural product without chemical additives; no toxic emissions during manufacture and installation; naturally resistant to moisture/mould due to high silicate content. It is CO$_2$ neutral.
Disadvantages:	Limited resources of raw material. Delivery distance from central Europe.

Application
Installation:	Cut boards with hand-held circular-saw, jig-saw or similar. Boards are then fixed in place with mineral adhesive and proprietary insulation fixings (walls) or galvanised screws with washers (ceilings and roofs). Profiled surface provides excellent key for plaster finish. Protect from moisture.
Durability:	Established building material with good history. When used externally protect from moisture (driving rain, splashing water, etc). Known life span of 70 years when installed as roof covering.
Removal:	Unplastered boards are easy to reuse due to mechanical fixing. It is not possible to reuse plastered boards.
Disposal:	Recyclable (wire). Compostable (reed). Incineration.

3.6.7 Wool/wool-rich insulation products

Manufacture

Sheep's wool insulation is made from hair lengths of around 4 to 55 mm. Raw wool (containing 50–70 per cent dust, salts and wool fat, etc) is scoured (washed), dried, carded, and then either woven or stitched into a roll or batt. Products may contain up to 18 per cent synthetic support fibres (non-recycleable compound).

Primary embodied energy: 15 kWh/m^3

Figure 3.11 *Wool insulation*

Areas of use

Thermal and acoustic insulation for internal walls and floors, and roofs; acoustic underlay for timber floors; as sealing strip between building components, ie around windows and doors.

Product details

Material:	Treated or untreated sheep's wool
Possible additives:	Synthetic fibres; mineral salts; sometimes larvacides and pesticides.
Form:	Batt, roll (quilt, felts, strips), loose (wadding)
Thickness:	40–250 mm (quilt), 50–100 (batts), 3–10 mm (felts)
Country of origin:	England, Germany, Switzerland Austria, New Zealand

Material properties

Product type:	Quilt	Batt	Felt	
Use classification:	W	W, WL	T, TK	(DIN 4108)
Thermal conductivity:	0.040	0.040	–	W/m K
Specific Heat capacity:	1720	1720	1720	J/m^2 K (wool)
Diffusion resistance factor:	μ=1	μ=1	μ=1	(DIN 4108)
Bulk density:	20	25	–	kg/m^3
Fire classification:	B2	B2	B2	(DIN 4102)

Ecological properties

Advantages:	Low energy during manufacture, non-irritant to skin, eyes or respiratory system. Wool is able to absorb between 20 and 40 per cent of its own weight in moisture without adversely affecting the thermal conductivity and the performance of the product.
Disadvantages:	Current UK production of wool insulation is able to cover only 0.1 per cent of the UK insulation market. Ideally the wool used in insulation products should come from sheep that have not been dipped to avoid the wool material containing chemical residue from the sheep dip. If dipping is necessary, then dipping should be undertaken well in advance of shearing. The synthetic fibres in wool-rich products do not have the same properties as wool and will release toxic emissions if burned.

Application

Installation:	The ability of moth larvae to access insulation can ideally be designed out by, for example, compartmentalisation, avoiding the need for treatment against larvae. To avoid slumping between the studwork, wool products may require tacking against the header plate.
Durability:	Products are generally stable, rot-proof, and durable. UK-produced wool insulation will remain effective as an insulant for the life of the building in which it is installed (according to BBA Certificate).
Removal:	Can be reused.
Disposal:	Products containing synthetic supporting fibres may require incineration.

4 Light structural wall materials

This chapter covers materials that serve a dual function in a building, being both part of the thermal insulation and a structural part of the wall. In general they are not carrying major structural loads, but are forming the body of the wall, and carrying their self-loads and wind loads. Usually they will be used with a timber-framed structure. The systems vary according to the proportion of the material used that is natural fibre as compared to the binder. Their method of construction also varies in terms of formed materials (hemp/lime) and blocks (light earth and straw bale).

Because straw bales use only straw they are presented separately, whilst the mixed fibre/binder products are brought together in one section.

There is an overlap with the use of fibres to limit cracking in plasters and some structural materials.

The technical data provided in this chapter is taken mainly from manufacturers' literature. It is therefore public domain information, but cannot be guaranteed, and the reader is required to check supplier data before use.

4.1 STRAW BALES

4.1.1 What are straw bales used for?

The use of straw bales in building is divided into two categories: structural (load bearing) or non-structural (in-fill). Both forms require breathable construction, so are ideally suited to construction materials and techniques discussed in this handbook. Straw is most commonly used in walls, but it has been used loose as a replacement for conventional cavity insulation in floor, roof, etc (although these uses have more serious fire and damp risks). It has also even been used as reinforcement for concrete in floor slabs etc.

For a comprehensive guide on the history and evolution of the straw bale form of construction, see (Swentzell, and Steen, 1994).

4.1.2 Why consider using straw bales?

Straw is the stalk of cereal crops, such as wheat, oats, rye and barley, left over from the harvesting process. Straw is used as bedding material for livestock, as mushroom compost or even thatch. Every year, approximately 4 million tonnes are produced surplus to requirement in the UK. This volume creates a major disposal problem since it cannot all be ploughed back into ground, and burning, which until recently was the most common disposal method, is now banned. Straw can be burned for energy generation, and this is a competing, but relatively low value use for the material.

Due to the lower weight of straw bale construction, the site impact and requirement for foundations is reduced.

Straw is most commonly available at relatively low cost and size as bales, which are easy to manhandle. Bales are easy to work with using conventional hand tools. This makes them ideal for use in self-build construction.

If designed and built correctly, no treatment to the straw should be needed.

4.1.3 Performance requirements for straw bales

In load-bearing construction, the structural integrity of the building is derived directly from the bulk of the straw bale, which must therefore be made with the straw well compacted and arranged in a rational compact manner. Bales should be made from freshly gathered straw – minimum length 150 mm, although 300–450 mm is preferable for greater strength. Each bale should be tied tight with a minimum of two strings, at least 100 mm in from the edges. The strings should be biodegradable (sisal or hemp) to avoid any disposal problems. Bales should be secured to the foundation, and to each other, usually by wooden stakes driven through several courses, in order to form a continuous element. The walls will be secured to the roof by continuous tie beams to prevent bowing.

For both load-bearing and in-fill constructions, bales should be kept dry and clean at all stages, from straw gathering through baling and transportation, and the construction process itself. Optimum moisture level is less than 15 per cent (wet weight basis) to minimise the risk of bacterial and fungal attack.

Requirements for in-fill construction are not as onerous since the integrity of the building is derived from the structural frame (usually timber, but can also be lightweight metal framing, etc). However, using bales with the optimum properties as given above will ensure a durable structure which will require the minimum of maintenance.

4.1.4 Performance data

These data are given in the data sheet in section 4.4 of this chapter.

4.1.5 Design guidance

The load bearing form is probably the most efficient use of straw but requires extra care in detailing to ensure a stable structure that does not settle unpredictably. Lime-based or similar natural rendering materials serve both to protect the straw from the elements and also add significantly to the strength of the structure. In practice the render can then be carrying much of the load, leading to risk of collapse if it is removed for replacement.

In-fill construction (timber frame) is relatively simple and does not differ greatly from more conventional construction methods. It is also easier to achieve clearance with building regulatory bodies.

The most important consideration in designing with straw is detailing, to protect against direct exposure to moisture while allowing the wall structures to breathe adequately. Protection from moisture is achieved by the use of a generous overhang on eaves (at least 500 mm) and physically raising the bottom course of bales to at least 200–300 mm above ground level (adopting a suspended floor construction, possibly using a concrete upstand below the first course). The wall structure can breathe properly by avoiding the use of vapour-proof membranes and similar gas-impermeable cladding material and using appropriate sheathing as shown in Figure 4.1.

Attention should be paid to protection against insects and rodents which can nest in the straw bale and degrade the strength and quality of the structure. This requires careful application of (lime) render or other coatings. The approach to plastering is to build up layers (Amazon Nails, 2001), starting with a thin layer applied by hand. The process of plastering then takes several days, as each layer needs time to harden.

Fire protection is also an issue, although it has been shown that the performance of straw bales is comparable to other typical building material when sheathed with conventional plasterboard or plaster render.

Electric cables can be encased in conduits to further reduce any risk due to overheating of cables, and these could be placed in the straw or in the plaster coating. Water pipes are best run within internal walls, as the effect of a leak would be significant.

The most common solution for the finish on straw is lime render. This is a traditional solution with hundreds of years of testing. It consists of lime and sharp sand, mixed together by hand or with a paddle mill as used by potters. The surface of the straw should be cut short and smooth, and the render built up in layers. The building outline shown as Figure 4.1 used a timber cladding, which is a less common solution, but still allows the straw to breath.

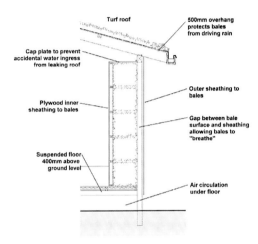

Figure 4.1   *Typical wall detail for a building with in-fill construction*

An interesting development for straw bale building is an attempt to produce larger, pre-manufactured straw panels. This was used on a new building for the University of the West of England at Bristol, and is shown in Figure 4.2 below.

Figure 4.2   *Large straw panel*

The reader is referred to the following publication/sources for more information, (Amazon Nails), (Swentzell, A and Steen B, 1994).

4.1.6   Applications

There are many examples of straw bale houses available on the Internet and in the publications referred to in this chapter.

Some typical examples are shown here

Figure 4.3   *Typical interior, with straw visible in one area at the Centre for Alternative Technology (CAT)*

Figure 4.4   *A small straw bale building at CAT*

Figure 4.5   *Straw being installed in a smokeshed*

Figure 4.6   *Completed interior*

4.2   REINFORCEMENT IN BLOCKS

4.2.1   What is the application for natural fibre in blocks?

Fibre has been added to blocks and plaster for thousands of years, in the form of cob walls, wattle and daub, reinforcement in sun-dried bricks and in plaster. It serves to strengthen the material and, in particular, to reduce cracking as the materials dry. There are no current mainstream uses of natural fibre in these ways, although the techniques remain in use for restoration and are widely used in less developed countries.

There is limited use of straw as reinforcement in unfired clay blocks, also known as "light earth". Unfired lightweight, hand-pressed earth bricks, made from a mixture of clay, sand and straw, can be used for non-load bearing walls. They have good acoustic properties and can help regulate temperature and humidity, making them ideal for "breathing" constructions. Typically a clay brick wall is used as a partition wall and is built using clay mortar before being covered with a clay or lime plaster.

4.2.2   Why consider using fibres?

There are several reasons to incorporate fibre in unfired blocks rather than blocks on their own:

- improving thermal performance
- increasing strength in use
- increasing strength during transportation
- crack resistance
- enhanced drying of blocks due to pathways for moisture to leave
- low embodied energy.

In traditional mud construction, the fibre is an essential part of the structure, but clearly this approach is little used in modern western construction. However, there is new interest in the use of unfired blocks to reduce the embodied energy of buildings, and this has re-awakened an interest in the role of fibres. Research at the architecture department at the University of Bath is looking into the UK production of unfired blocks. Light clay blocks are available from Germany, and these provide for good thermal stability, acoustic performance and moisture control.

4.2.3   Performance requirements for reinforcing fibre

Because the role of the fibre varies between the different applications it is difficult to be precise about the performance needed. Fibres need to be clean and reasonably dry to help the bond with the other material. The percentage of fibre used is relatively low with 4 per cent by volume being typical for high strength materials, although higher percentages can be used for more insulation.

4.2.4   Performance data

As discussed above, there is only limited precise technical data in what is a varied field of activity. Further it is clear that the properties of a particular pairing of materials will vary with the proportions of the two materials and their individual properties. It is therefore difficult to give any more than indicative performance data.

Unfired blocks

Unfired straw/clay bricks are useful as thermally insulating building materials and have a thermal conductivity of 0.21 W/mK, as compared to 0.95 W/mK for solid clay. Their density at around 700 kg/m^3 is close to half that of solid clay blocks.

Figure 4.7   *Solid Clay blocks*   **Figure 4.8**   *Light clay blocks*

4.2.5   Design issues

A good source of guidance on the use of fibres in plaster and blocks is *Appropriate Building Materials* (Stulz and Mukerji, 1993) prepared by experts in "intermediate technology" suitable for low-cost construction in developing countries.

For reinforcement of clay blocks, the fibres are recommended to be dry, cut to approximately 6 cm in length and well mixed to avoid insect nests. A ratio of around 4 per cent by volume is found to be appropriate, when this is much higher the strength decreases. It is possible to use most sources of fibre including wheat, rye, barley, sisal, hemp etc.

A recent DTI funded research project has developed further guidance on the use of "light earth" in construction. The project has produced a website, <www.lightearth.co.uk> that contains a number of case studies from the UK and overseas, and information on thermal and fire performance.

4.2.6   Applications

There are full examples of light earth construction included in the Chapter 9: Case studies.

4.3   HEMP AND LIME FIBRES IN CAST CONSTRUCTIONS

4.3.1   What is the application for hemp and lime?

Hemp and lime, combined together, make a material that can be formed into a wall or floor material, suitable for interior and exterior surfaces. In principle it could be formed in blocks, but it seems to offer most potential as a formed or cast material. This is then similar in use to concrete, but with better thermal and environmental performance.

4.3.2   Why consider hemp and lime?

There are a number of potential benefits from using hemp and lime:

- lower embodied energy construction than brick or concrete
- single process of construction for wall and insulation
- good synergy with timber frame building
- reduced heating energy use
- buffering of indoor moisture
- good quality internal environment.

It should be noted that the use of hemp with lime is a new technology, and there is only a small number of examples built in the UK. There are more that have been built in France and other countries. The houses at Haverhill, featured in the case study in Chapter 9, use this method and the case study provides further information on the materials.

4.3.3   Design issues

Because the construction is carried out in a different way from "conventional" building, the decision to use this technique affects the whole of the rest of the design. Further advice from the materials suppliers, or those with experience of the material, should be sought at the earliest stage of design.

Although in principle the hemp and lime can be load bearing, it is usually built with a timber frame system, so that only self-load and wind loads are being carried. It is hard to quantify the performance of these materials, because these vary with the precise mixture. It is reasonable to note that:

- the combination gives reasonable thermal performance from structure
- the lime makes it insect and rot resistant
- the lime makes it fire and water resistant.

There is current research in progress as part of a planned new development at the Centre for Alternative Technology, to improve the understanding of the performance of hemp lime.

4.4 DATA SHEETS

4.4.1 Straw bale construction

Manufacture

The straw used in straw bale construction is typically obtained from cereal crops such as wheat, oats, rye, barley and maize. Straw bales are made from the pressed stalks that are left when the head of the crop is removed. The straw is harvested, and then compressed into approximately 100 mm thick layers that are stacked to form bales. Bales are usually tied together with polypropylene string.

Primary embodied energy: 0.13 MJ/kg

Figure 4.9 *Straw bale house*

Areas of use

Straw bales are typically used as thermal insulation for external walls, either load-bearing or, more commonly, as infill between a structural frame. Straw bales can be used as thermal insulation in roof constructions and in ground floor slabs.

Product details

Material:	Straw (cellulose, lignin, silicates, hydrophobic wax coating)
Additives:	Generally none (sometimes waterglass, borax, boric acid)
Form:	Rectilinear bales
Thickness (wall):	320–340 mm (bales) finished with 25 mm plaster to both sides.
Country of origin:	Local to use

Material properties

Product type:	Minimum	Typical	Maximum	
Bale format:	500 x 400	500 x 750	500 x 1200	mm
Thermal conductivity:	0.055	0.060	0.065	W/m K
Specific Heat capacity:		–		J/m^2 K
Diffusion resistance factor:		$\mu = 5$		(DIN 4108)
Bulk density:	112	133	–	kg/m^3
Moisture content	5	14	15	%
Compressive strength:		41–69		N/m^2
Fire classification (bales):		B3 / F30		(DIN 4102)

Ecological properties

Advantages:	Straw is an abundant material that is available throughout the world because straw is a major by-product of the food/grain industry. The tube-like structure of the straw gives straw bale constructions both good elastic and thermal performance. The environmental impact of the straw is negligible as it is completely biodegradable.
Disadvantages:	Settlement, durability

Application

Installation:	Relatively simple and quick due to the size of the bales.
Durability:	Modern harvesting methods are able to remove almost all of the head (or ear) of the plant from the straw resulting in an insulation material that is of little interest to rodents and bacteria. Straw bale construction is dependent on appropriate detailing to provide durability.
Removal:	Easily removable at end of building life.
Disposal:	Untreated straw will slowly biodegrade.

4.4.2   Light earth (straw-clay) construction

Manufacture

Light earth construction is a generic name for constructions made with earth mixed
with either an organic or mineral lightweight aggregate. Organic fill materials are
typically straw, woodchip, cork granules, wood shavings, or sawdust. The straw
used in light earth construction is typically obtained from cereal crops such as
wheat, oats, rye and barley but the stems of flax, hemp, and reed can also be
used. The straw should have a moisture content less than 16 per cent and stalk
length 5 or 6 cm.

Figure 4.10 *Light earth wall*

Areas of use

Thermally insulating monolithic infill between structural frame for external and internal walls;
thermally insulating prefabricated light earth elements, usually blocks for infill construction to
external walls and non-load bearing internal walls and inner wall linings.

Product details

Material (straw-clay):	Lime (clay, silt, sand), lightweight aggregate (straw), water
Typical additives:	Small plant fibres, sawdust, rye flour, clay mortar, cellulose adhesive
Form:	In-situ – loose mix supplied in bags. Modular blocks – supplied on pallets.
Thickness (wall):	100–300 mm (in-situ) finished with 25 mm render to both sides.
Country of origin:	Local to use (wall mix). Germany, Austria, Switzerland (blocks).

Material properties

Product type:	Wall Mix	Block A	Block B	
Format:	Cast in-situ	500 x 250	115 x 240	mm
Thermal conductivity:	0.09 - 0.47	0.14	0.21 (0.23)	W/m K
Specific Heat capacity:	1300–1000	1200	–	J/m^2 K
Diffusion resistance factor:	μ = 4–10	μ = 4–6	μ = 4 (4)	(DIN 4108)
Bulk density:	300–1200	1000	700 (750)	kg/m^3
Moisture content	2.5–4.5	–	–	%
Compressive strength:	2–5	2.3	–	N/ mm^2
Fire classification:	B2–1/F30–90	B2/F90	–	(DIN 4102)

Ecological properties

Advantages:	Straw and clay are both abundant materials that are available throughout the world. Straw and clay work well together, the straw provides the thermal insulation, the clay binds the straw together and protects it from excessive moisture. The environmental impact of a clay-straw is negligible if materials are locally sourced.
Disadvantages:	Wet mixes require shuttering which can be labour intensive.

Application

Installation:	Dependent on the construction methods employed.
Durability:	Insured life of 30 years for light earth walls comprising "organic" straw fill material – clay mix cast in-situ around timber frame structure.
Removal:	Easy to remove at end of building life.
Disposal:	A straw and clay mix can be dug into the ground and left to decompose.

4.4.3 Light lime (hemp-lime) construction

Manufacture

Light lime construction is a generic name for monolithic elements made from lime mixed with a lightweight mineral or mineralised aggregate, typically mineralised hemp shives. Sand can also be added to the mix. Hemp shives, typically 20–25 mm in length, are a by-product of the mechanical process which removes hemp fibre from hemp stems. There is a process, called petrification, used by some producers to improve the resistance of the hemp to fire and rodents.

Figure 4.11 *Hemp lime interior*

Primary embodied energy: -1.15 kW/m^2

Areas of use
Thermally insulating monolithic structural walls and floors (no sand); thermally insulating monolithic solid floors (containing sand); below sand and hydraulic lime screed.

Product details
Material (hemp-lime):	Hydraulic or hydrated lime, hemp shives, water
Typical additives:	Sand
Form:	In-situ – hemp supplied loose in paper sacks; lime supplied in containers
Thickness (wall):	150–300 mm (in-situ) finished with 10 mm lime render to both sides
Country of origin:	France

Material properties
Product type:	Wall Mix (no sand)	Floor Mix (with sand)	
Format:	Cast in-situ	Cast in-situ	
Thermal conductivity:	0.12	–	W/m K
Specific Heat capacity:	–	–	J/m^2 K
Diffusion resistance factor:	–	–	(DIN 4108)
Bulk density:	550	850	kg/m^3
Moisture content	–	–	%
Compressive strength:	0.458	0.836	MPa
Fire classification:	M1	M1	

Ecological properties
Advantages:	The growing of hemp does not require the use of herbicides, fertilisers or pesticides.
Disadvantages:	Requires shuttering which can be labour intensive.

Application
Installation:	Cast into shuttering as wet conglomerate and allowed to carbonate and dry out.
Durability:	The BRE concluded that the durability the hemp houses at Haverhill, Suffolk, was at least equal to that of traditional constructions.
Removal:	Easy to remove and break up at end of building life.
Disposal:	Hemp-lime aggregate can be used as a soil conditioner.

5   Paints and finishes

5.1   WHAT ARE PAINTS AND FINISHES USED FOR?

Note that this chapter is different from the other chapters because although there are paints that contain crop-based materials, they are not as distinct from each other as the other products are. Formulations by different manufacturers will contain different ingredients, and it is therefore not appropriate to include data sheets.

Paints and finishes have the following main roles in construction:

- protecting materials from moisture, dirt, dust, grease etc.

- enhancing the aesthetic value of building interiors and exteriors

- changing the visual appearance of surfaces and objects

- creating a comfortable living and working environment for building occupants.

The variety of surfaces to be coated are also subject to different environmental conditions and wear, which lead to many variations of paints and finishes.

Figures 5.1 and 5.2 below illustrate the range of natural-based paints and colours available on the market at the time of writing.

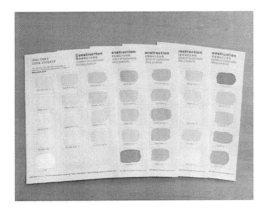

Figure 5.1 *Natural paint colours*   **Figure 5.2** *Assorted natural paints*

5.2   WHY CONSIDER CROP-BASED PAINTS?

The potential benefits of crop-based paints and natural ingredients used are:

- special performance characteristics

    - flexibility – avoids cracking when the surface expands and contracts

    - porosity – enables moisture to pass through

    - ease of maintenance – breathing finishes can easily be patched

    - historic authenticity – for restoration work.

- minimal environmental impact
  - low embodied energy
  - low pollution – crop-based paints provide a much lower release of volatile organic compounds (VOCs), which have been implicated in sick building syndrome, plus respiratory and neurological heath effects (Wooley 2002).
- maintaining optimum indoor air quality
- low or non-toxic materials
- reduced disposal issues.

All paints and finishes, whether natural or synthetic, contain a more or less complex mixture of ingredients, which together provide the performance required of the product. Natural paint manufacturers do everything possible to keep the composition of their products as simple as possible and to use crop-based and other natural ingredients wherever these are appropriate.

A distinguishing feature of natural paint manufacturers is that they normally declare all the ingredients of their products – information not normally found on paints from the chemical industry. This allows people with particular sensitivities to check whether the product might affect their health, whether they are users or occupiers.

Crop-based materials are used both in natural paints and in some synthetic paints, but often in relatively small proportions to the total volume. Mineral and timber-based materials, and of course water or organic solvents, often form the major proportion of content by volume, and this applies to both natural and synthetic paints.

A wide range of colours and paint types is available from natural sources to meet most needs. The most challenging condition is where the finish is permanently exposed to water, for example on timber in damp soil. It should be noted, however, that not all natural materials are environmentally neutral or non-toxic. Plants that can have damaging side effects abound in nature. Their successful use, as with all materials – natural or synthetic, requires good knowledge of their properties.

5.3 PAINTS AND FINISHES COMPONENTS FROM CROPS

Agricultural crops provide a significant proportion of the content of natural paints and finishes, but most commercial "non-natural" paints use a combination of mineral-based and synthetic ingredients, of which a large proportion are petrochemical-based.

Even in natural paints, crop-based products are only part of the mixture. Mineral products (such as clay, chalk, talcum and marble powder) and timber derivatives (such as cellulose, colophony, dammar and citrus terpene) make up a significant proportion of the content. The majority of ingredients in most natural oils and waxes for floors and other timber surfaces are tree-based and, as such, are outside the scope of this study.

The ingredients of paint have a variety of functions:

- **pigments** provide colour, and constitute typically about 30 per cent of paint content
- **binders**, typically around 34 per cent of content, carry the colour during application and make it adhere to the surface when dry

- **solvents** and emulsifiers, also about 34 per cent of content, are mixed in to produce a good working consistency

- **drying agents** and other additives such as plasticisers, preservatives and fungicides, are added in small amounts, typically 2 per cent of content, to make the product dry within a reasonable time and to improve other aspects of the paint's storage and performance

- **fillers** may be added to pigments to bulk out the consistency and to improve coverage.

Tables 5.1 and 5.2 below list the potential plant-based ingredients that can be found in different compositions of paint, grouped according to the role they play in the mixture. These are followed by a brief description of key properties of the more significant examples. The list below includes plants which were cultivated in the past to provide pigments, but are not currently cultivated in significant quantity. Natural paint manufacturers in Germany are actively investigating the possibility of re-starting large-scale cultivation of some of these crops.

Table 5.1   *Pigments and dyes from plants*

Pigments and dyes	Colour
Madder root (*Rubia tinctorum*)	red (Turkish red)
Reseda or weld (*Reseda inteola*)	yellow to olive green
Wild woad (*Isatis tinctoria*)	blue
Cultivated woad	yellow
Indigo (*Indigofera anil & tinctoria*)	blue
Various vegetables	various

Figure 5.3   *Assorted natural pigments*

Table 5.2 *Other paint elements from crops*

Binders and thinners	Solvents and emulsifiers	Additives
Linseed oil (*Linum usitatissimum*) – extracted from flax seeds	Citrus peel oil (D-Limonene *citrus terpene*) – extracted from orange peel by cold pressing and steam distillation	Rosemary oil – mild fungicide and preservative
Soya bean oil (alkyd resin)	Soya (source of lecithin, used for solvents and emulsifiers)	
Sunflower oil (*Helianthus annuus*) – extracted from sunflower seeds	Maize (source of lecithin)	
Poppy seed oil	Peanuts (source of lecithin)	
Safflower oil (*Carthamus tinctorius*)	Egg	
Casein (dried milk curd with strong adhesive qualities, used both as a base for high-performance, soft white wall paint and in adhesives)	Vinegar ester (made by distilling vinegar.	
Plant alcohol (ethanol)		
Vinegar ester		
Egg		

Other crop-based components include:

- **olive oil:** mixed with vegetable oils, alkaline ash from sea plants and Mediterranean salt water to produce Marseilles soap.

- **palm oil:** used as an alternative to olive oil in Marseilles soap.

- **beeswax:** a supple but strong wax, produced by bees to form their honeycombs, and used for centuries on furniture and other wooden surfaces to provide a fragrant protective sheen. The raw material is 100 per cent natural, but many commercial beeswax polishes include a variety of chemical components.

- **vegetable dyes:** a variety of vegetables and plants can be used to produce dyes. Because light can pass through the molecules of colour, these dyes have an attractive translucent quality. But plant dyes are not light-fast and tend to fade over time. Different plants produce different colours, eg beetroot: red/pink, onion skins: yellow, blackberries: purple.

- **vegetable oils:** used in the production of Marseilles soap.

Available natural finishes

Natural finishes are available commercially for the applications listed in Table 5.3 below.

Table 5.3: *Available applications of natural finishes*

Walls and ceilings	Floors	Internal timber	External timber
Natural emulsion	Primer	Primer	As for internal timber, but also:
Casein wall paint	Hard oil	Undercoat	• UV-resistant varnish
Clay finish plasters	Floor oil	Gloss and eggshell finishes	• "Scandinavian red" and black for cladding
Clay paints	Varnish	Hard oil	• Fence paint
Silicate paints	Stain/lasure Marseilles soap	Stain/lasure	

Many of the raw materials used today (particularly minerals) are common to both natural and synthetic paints. It is mostly natural paint manufacturers who choose to use crop-based materials as much as possible, despite their often higher cost. They do so to be able to produce products of the highest quality, with many long-term performance advantages. Further they seek to avoid damage to the environment and protect both painters and building occupants from the potential health hazards of the complex chemical combinations in synthetic paint formulations. Environmental damage from paints and finishes can arise from obtaining the raw materials, from the manufacturing process and from eventual disposal. Some examples include:

- For many decades massive volumes of the toxic "cake" resulting from the production of white pigment from titanium dioxide, mostly in Germany, were simply dumped in the North Sea (now stopped).

- The manufacture of 1 kilogram of petrochemical tar pigments such as signal red, produces 300 litres of polluted water .

- As an indication of the level of chemical complexity involved in the paint industry, about one third of all new chemical compounds developed are pigments.

Some examples of natural paints and finishes and their properties

Natural emulsion

- close equivalent to synthetic emulsion in appearance and use, on walls and ceilings

- emulsified linseed oil in water or natural latex used as binding medium

- highly stable and scrubbable finish

- less breathing than lime wash or distempers

- biodegradable

- non-toxic

- can be tinted with earth or mineral pigments.

Casein wall paint (an alternative to emulsion)

- casein (milk protein) mixed with clays and chalk

- high opacity

- anti-static: does not attract and retain dirt particles

- breathable

- durable and very stable: ages more slowly than other paints

- attractive soft chalky finish

- easy to apply

- fast drying

- inexpensive

- contains no solvents

- 100 per cent natural + low energy use in production

- 100 per cent biodegradable/compostable

- supplied in powder form for mixing with water, reducing transport energy costs

- ideal for historic renovation work.

Linseed-oil and linseed-oil based finishes

- made from flax seeds

- highly penetrative: extremely small molecules, 50 times smaller than those in synthetic binders, allows deep penetration of timber pores, making linseed oil much more effective than synthetic resins

- highly flexible: long drying time retains flexibility

- vapour permeable

- use waste products

- renewable resource

- completely non-toxic

- high quality finish

- natural paint manufacturers use 100 per cent organic linseed oil.

Figures 5.4 and 5.5 illustrates examples of natural paints in use

Figure 5.4   *Example of natural paint 1*   **Figure 5.5**   *Example of natural paint 2*

5.4   PERFORMANCE, DESIGN AND SELECTION ISSUES

The applications of paints and finishes are diverse and consumers' choices are governed by a wide range of factors, many of them subjective rather than objective. Table 5.4 below shows the main criteria to which manufacturers will attach different importance, alongside cost, in determining the mix of ingredients for their products.

Table 5.4   *Some determining manufacturers' choices of ingredients for paint*

Fashion	Intensity	Translucence
Texture	Softness to touch	Hardness/non-scuffing
Coverage	Ease of application	Absence of drips
Durability	Ease of maintenance	Washability
Speed of drying	UV resistance	Smell
Opacity	Colour-fastness	Breathability

In general there will be a natural paint or finish for most circumstances. However, the performance criteria and the breakdown of the main functional constituents of paint in Table 5.4 above show why it is not possible to give a simple answer to questions about selection. Each manufacturer selects the combination of crop-based and other materials judged to produce products that best match their view of market requirements.

Each user and specifier will need to make individual decisions about balancing cost and performance with both personal and wider environmental considerations. For example whether to:

- choose a product with a strong-smelling but completely natural citrus oil solvent or one that uses a neutral-smelling but petrochemical-based paraffin derivative

- accept the range of colours and intensities achievable with earth or mineral-based pigments or to use a certain percentage of synthetic pigment to achieve colours perceived as stronger or brighter

- use beautiful, intense, but very expensive natural ultramarine blue, or to settle for much cheaper synthetic blue: the synthetic blue will be hard to distinguish visually, and it will probably not be directly toxic to painter or occupant, but the embodied contribution to wider environmental damage may be significant.

To get the desired outcome it is therefore necessary to specify the product by name rather than by generic type.

The use of the surfaces to be covered and their exposure to weather are crucial in

(a) determining which product/s to use and

(b) establishing what cyclical maintenance will be required.

What is critical to the long-term success of any finish is that a detailed maintenance schedule should be drawn up and included in the specification. In much of Europe, conventional paint manufacturers and, increasingly natural paint manufacturers, will entertain claims about the long-term performance of their products only if appropriate maintenance contracts have been implemented between painter and client, covering six-monthly or annual inspection for damage, patching up and regular re-coating. Even for powder-coated aluminium-clad windows, normally thought of as extremely durable and low-maintenance, manufacturers recommend cleaning every three months in built-up areas.

Some of the most challenging conditions for finishes on timber are the combined presence of moisture and air in soil. Only the most environmentally aggressive chemicals (like creosote) have a long lifetime in these conditions, so it is not reasonable to expect a natural product to last forever. It is better to select a durable timber, from a sustainable source rather than rely on treatments.

Drying times

An important factor to bear in mind with natural paints and finishes is that they typically take longer to dry than synthetic products. This is because many of the chemicals in paints are there to help them to dry quickly, and use of these agents should be kept to a minimum to reduce environmental impacts. So it is generally necessary to plan for longer drying times when using natural paints, and accept this as a consequence of the choice of a less environmentally damaging material.

Colour range

A further issue relates to the range of available colours. It is not possible to produce every possible colour and hue through natural or crop-based materials: very bright reds are one example. In addition, natural materials for some colours are extremely expensive: ultramarine blue is perhaps the most common example, with synthetic blue pigment costing a fraction of the very high price of mineral-based natural blue.

Repeatability

Many natural raw materials are inherently variable in their consistency and colour, and paints made using them are more likely to show slight variations from batch to batch than standard products from a chemical production line. These small variations can bring a special quality and life to surfaces, but care should be taken where complete consistency of hue is wanted.

Applying the finishes

Natural paints and finishes are generally applied in the same way as conventional finishes. However, there are sometimes important differences in relation to the tools and equipment that should be used and such things as the number of coats to use, the thickness of each coat, direction of application, etc. Successful application of natural paints requires understanding of these differences.

5.5   APPLICATION

5.5.1   Office refurbishment, London.

What

A London-based firm of consulting engineers wanted the refurbishment of their offices to have as little impact on the working environment as possible. It wanted to minimise the disruption to the office caused by the off-gassing of newly-painted surfaces and aimed to create a comfortable and healthy indoor environment. Indoor air quality was an important issue for the client, so wholly natural interior finishes were specified.

Materials

The specification required for the use of natural plant-based interior paints and finishes was linseed oil based emulsion, gloss paints and timber finishes. These products did not include synthetic binders, pigments or other compounds.

Figure 5.6   *Office view 1*

Figure 5.7   *Office view 2*

6 Crop-based products used in floor coverings

Agricultural products form the basis of important product ranges in the floor coverings industry, specifically linoleum and natural-fibre carpets.

Unlike natural paint manufacture, which remains relatively small scale, linoleum and carpet manufacture are mainstream industrial activities and the raw materials involved are required in large quantities. Given the similarities between many of the products, there are no data sheets provided in this section.

6.1 WHAT ARE FLOOR COVERINGS USED FOR?

Floor coverings serve multiple purposes in a building:

- they cover rough, raw surfaces such as concrete and timber to provide a comfortable surface on which to walk
- they form one element of the aesthetic appearance of a room
- they can improve the thermal insulation of the floor
- they can improve the acoustic performance of the room
- they can improve the grip to provide a safer flooring material.

6.2 WHY CONSIDER CROP-BASED FLOOR COVERINGS?

Floor coverings are much simpler products than paints and finishes, and it is possible to buy floor coverings that are 100 per cent natural. These are desirable for several reasons:

- to avoid health hazards
- for greater comfort
- for better performance
- to minimise use of non-renewable raw materials such as oil
- to avoid accumulating unmanageable volumes of non-biodegradable waste.

Health issues

Nearly a third of all 13 to 14 year-olds in Britain (*The Guardian*, 2004) are affected by asthma, giving Britain the third highest incidence of teenage asthma cases in the world, and the highest in Europe – and the numbers are still growing. Children are missing school and being hospitalised, and similar proportions of adults are forced off work.

Allergies, triggered by dust mites, are believed to cause 80 per cent of asthma cases (Net Doctor, 2004). The causes are insufficiently understood, but the increasing gravity of the epidemic lies behind a five-year European-wide research programme launched in February 2004 with €14.4 m of European Commission funds. What is already clear is that the increasingly complex cocktails of chemicals to which we are exposed in all aspects of our daily lives are at least one element of the problem, and carpets have for some time been identified as a source of health hazards. Most people spend a significant proportion of their time on carpets, and small children are the closest to them. The carpets provide a home for the dust mites and, even after vacuum cleaning, a large proportion of the mites will remain in the carpet, unlike on linoleum and timber floors.

Artificial fibres and sheet materials such as PVC-based vinyl use a variety of synthetic components, but even when natural material such as wool is used for the pile of carpets, this is usually contaminated with pyrethrin or nerve-affecting moth-proofing agents (actually required to obtain the Wool Mark label). Further, the backing of most carpets, whether made with natural or artificial fibres, is generally described simply as "latex". It is not natural latex, but usually a form of styrene-butadiene-rubber (SBR). SBR is based on non-renewable petrochemicals, and contains a cocktail of chemicals such as stabilisers, fire-retardants, vulcanising agents and softeners. Studies have shown many of these chemicals, and their combinations, to be health hazards.

Floor coverings based on crop-based materials avoid these problems where chemical treatments can be avoided. Wood-fibre boards, flax or wool felt, natural or recycled rubber, or corks provide good alternatives as underlay material.

Greater comfort

Synthetic fibres and plastics have little or no ability to absorb moisture. This makes them badly suited to internal environments where relative humidity levels fluctuate as a result both of the external climate and of the occupants' presence and activities such as cooking. The health and comfort of the occupants depend on both temperature and humidity remaining within a certain "comfort zone" – for humidity generally between 40 and 55% RH.

Floor coverings based on crop-based materials are able to work alongside other breathing elements in the room, such as walls and ceilings, to maintain a comfortable living and working indoor environment, by temporarily absorbing swings in both humidity and temperature. However where spills of liquids are likely to occur then a waterproof option will be needed.

Better performance

Synthetic floor coverings are notorious for building up electrostatic charges. The problem of walking down a carpeted corridor then getting a sharp shock when touching a metal doorknob is less common now than in the 50s and 60s, but it does still happen. Instructions for handling electronic components still put emphasis on avoiding static discharges.

This characteristic of artificial floor coverings also affects their performance by attracting and retaining dust and dirt: keeping them clean is more of an issue. Shrinkage in PVC floor coverings as plasticisers evaporate, can produce damaged joints which turn into dirt traps and a breeding ground for fungus. Plastic surfaces do not absorb moisture, leaving the moisture more available for mould and bacterial growth.

Floor coverings based on crop-based materials are naturally anti-static: coir matting, for example, is often used on the floor of computer rooms for precisely this reason. Natural floor coverings are generally easier both to clean and to keep clean than artificial materials: sea grass matting is particularly good at shrugging off stains and dirt. Cleaning can – and should – be done using water and simple natural cleaners rather than aggressive chemical compounds.

Minimising use of non-renewable raw materials and energy

Synthetic fibres such as polyamide, polypropylene and polyacryl are all based on non-renewable oil, and also use significant amounts of energy in their manufacture. This also applies to most common carpet underlays and particularly to the synthetic latex, SBR.

Floor coverings based on crop-based materials are by definition renewable, and all use very much less energy in manufacture than the artificial alternatives.

Biodegradability

Britain's buildings contain a huge acreage of carpeting in a variety of forms, and these are typically renewed a number of times over each building's lifetime. Altogether this amounts to a massive volume of waste each year. The overwhelming majority of this carpet waste is not biodegradable and has to be disposed of either in landfill or in waste incineration plants. Some carpet should be recyclable, but for the main part it is not recycled.

6.3   PRINCIPAL FLOOR COVERINGS USING CROP-BASED MATERIALS

6.3.1   Linoleum

Linseed oil, a principal constituent of linoleum, is also used in paints. Linoleum contains about 23 per cent of linseed oil by weight. The oil is boiled with a drying agent before being oxidised then mixed with the other constituents.

Linoleum was first manufactured in England in 1864. It came to be used very extensively for domestic, commercial and institutional floors, but became less common after the Second World War, with the rise of vinyl tiles and other petrochemical-based floor coverings. It has since enjoyed a considerable renaissance and is rightly valued as a moderately-priced, highly durable and easy to maintain material that can provide a very attractive floor covering – while remaining very environmentally friendly.

Figure 6.1 *Sisal rug on lino flooring*

Figure 6.2 *Jute samples*

The main constituents of linoleum are linseed oil (23 per cent) and wood flour (30 per cent). Other components include softwood resin (8 per cent), cork flour (5 per cent), limestone powder (18 per cent), pigments (4 per cent), jute backing (11 per cent), and a drying agent (1 per cent, usually zinc). Thus most of the constituents are from renewable resources, and normally 100 per cent biodegradable. This compostability can be affected by the addition of plastic-based surface coatings, by the most commonly-used ethylene vinyl acetate (EVA) glue or if toxic colour pigments are used. However, glues based on natural latex can provide a completely satisfactory alternative to EVA glue.

6.3.2   Woollen carpets

Wool has been used in carpets for many years. The quality of woollen carpets varies enormously, depending on type and quality of manufacture, the quality of the raw wool, and the dyes and patterns used.

One manufacturer of carpets and carpet tiles uses goat's hair as the principal fibre. Artificial materials are however used for other elements of these products.

Wool is particularly susceptible to infestation and subsequent damage by moths – to the point of total destruction. This is a frequently-occurring problem even with clothes stored in cupboards but nevertheless moved from time to time. The problem becomes accentuated when sections of carpet get covered by infrequently-moved wardrobes and other items of furniture, creating ideal dark and undisturbed homes for moths.

The standard "solution" to this problem is the use of aggressive chemicals such as permethrin. Permethrin is judged to be carcinogenic and banned in the USA. It should certainly in any case be avoided by the growing number of people suffering from acute chemical sensitivity.

Most carpets prominently advertised as "100 per cent pure wool" not only contain permethrin or other anti-moth agents, but in most cases also contain SBR artificial latex in their construction (see section 6.2). Some chemical dyes can also represent a health hazard.

It is possible to buy woollen carpets that are made from 100 per cent untreated wool, with completely natural materials such as jute or hemp used as backing. There are also plant-based and non-toxic chemical products available both for moth-proofing and for cleaning such carpets.

Figure 6.3 *Wool carpet samples*

Woollen carpets are manufactured in a wide range of density and thickness of pile, using different methods to fix the pile to a variety of backing materials. The quality of the wool used, the quality of the backing, the method of production, and the density and thickness of the pile all play a part in producing carpets of widely varying durability and comfort underfoot.

In the Middle East and Asia, locally-sheared wool was traditionally coloured with plant-based dyes and woven by hand on domestic looms. Much hand weaving still survives, but in most cases synthetic dyes have replaced the original natural dyes. Industrial production has become much more prevalent in these countries and has taken over completely in Europe and the Americas, except for individual craftspeople who still hand-weave on looms.

Earlier industrial manufacture of woven woollen carpets such as Axminster, Wilton and Kidderminster used the natural backing materials such as hemp and jute that were the only choices at the time. The quality varied, of course, and not just in the thickness and density of the pile. Synthetic dyes were already being used but, in general, these carpets were more durable and less environmentally problematic than much of what followed.

6.3.3 Jute

Jute is used in large quantities as the backing for linoleum. However, soft, fine jute fibre also makes very attractive carpeting for lighter domestic use. It was traditionally used as a backing material for woollen rugs, and is still used today as a backing for other natural floor coverings, including coir and sisal.

It comes from a plant, genus *Corchorus*, grown in hot, humid areas of Asia, mainly Bangladesh and India. Extracting the long fibres from the plant's leaves for spinning into yarn is a very labour-intensive process. Its production provides rural employment for some 6 million people and in some cases is the only source of income in some of the world's poorest areas.

6.3.4 Sisal

Sisal, from the leaves of the *Agave sisalana* provides a tough and very durable natural floor covering. It is suitable for quite heavy wear as a carpet and is quite often used in commercial and public spaces. The raw material comes from tropical Africa, Central and South America, and Asia (mainly China), where it provides rural employment.

Sisal produces quite a rough, hard surface that many people find uncomfortable for bare feet and that would not be suitable for a baby's room. The material is quite inflexible and therefore not easy to cut straight. Its resilient structure also means that cut edges need to be "whipped" or bound with a linen edging strip to prevent fraying.

The appearance of sisal varies from country to country. For example, African sisal tends to have a shinier surface than Brazilian sisal. As well as its natural colour, sisal can be dyed, in a somewhat restricted variety of colours, in its country of origin. An example of sisal flooring is given as part of figure 6.1 above.

Most sisal carpeting sold in Europe suffers from the same problem as most ready-backed wool and other carpets: the backing material is synthetic latex. One manufacturer in Europe applies 100 per cent natural latex backing to the sisal carpeting which they buy from Brazil and Africa.

6.3.5 Coir/coconut fibre

Fibre from the coconut shell is woven, after a lengthy preparation process, to make very tough carpeting. The fibre is extremely resilient, anti-bacterial and non-rotting, and makes an excellent, fully-breathing component in mattresses and upholstery. It has very good acoustic insulation properties. In a denser mat form it is suitable for insulation under screeds, and it is used in the form of sound-insulating panels in the car industry.

It is produced mostly in Asia, in countries such as India and Sri Lanka, and the production process is quite long and still very labour-intensive, despite some moves to mechanise the process.

In Asia it is used in a wide variety of products, but in Europe today it is mostly known in the form of matting. The matting is very resilient and tough, but cut edges need to be bound to avoid fraying. In the UK it is most often available either bleached or in its natural, undyed state, but dyeing in the countries of origin is common. Bright colours may still be achieved using cationic dyes, which are banned in the UK because of their toxic qualities.

Coir is very hygroscopic, and will help maintain a comfortable indoor climate by absorbing swings in humidity. The downside of this property is that it tends to be dimensionally unstable particularly if it gets wet, when it may shrink unevenly on drying out. It is anti-bacterial and naturally anti-static, and for this reason it is often used in computer rooms. It will fade if exposed to direct sunlight.

6.3.6 Sea grass

Sea grass comes from China, and makes a tough, attractive floor covering. As the name suggests, it is a grass, woven into yarn for weaving. Its impermeability makes it hard to dye by conventional processes.

Sea grass provides a tough, stain and dirt-resistant and totally anti-static floor covering suitable for light to medium contract use in spaces such as galleries, restaurants, night-clubs and offices – although it is not suitable for areas with heavy footfall or without additional covering under caster chairs. It is supplied mostly in its natural colour, a beige/yellow mix, with hints of russet and green, and in a variety of weaves including herringbone and basket-weave.

6.3.7 Rush matting

Rushes are grown as a crop in many countries, including the UK. Rush matting has been used on floors in Britain since mediaeval times – and is still used to a small extent today. Flattened rushes are plaited and sewn into individual squares which are in turn sewn together to make mats of varying sizes.

Rush matting can be a very attractive floor covering on its own, but was also used in the past as an underlay for carpeting. It is much softer than, for example, sisal or coconut matting, and correspondingly gentle underfoot, but the weave is much more open, and the matting less durable.

6.3.8 Bamboo

Bamboo is an exceptionally fast-growing species of grass, rather than a tree. It is very widely used in Asia, both as a structural building material and for scaffolding, as well as for flooring and countless smaller products.

One species of bamboo in particular, *phyllostachys pubescens*, is used to produce a very hard-wearing floor material by a process of cutting strips of bamboo and laminating them into flat boards in a similar way to timber floors.

The resulting flooring panels are very robust. The bamboo is harder than oak or beech, making these panels suitable for demanding home and office use and for public spaces such as canteens, reception halls, and night clubs.

As with timber laminates, the environmental performance of the product depends significantly on the adhesives used, both within the laminate itself and to fix it to a concrete screed. The best-known European brand of bamboo flooring, available in a variety of colours, patterns and finishes, uses a ureum formaldehyde glue which meets the European E1 standard for formaldehyde emissions.

Figure 6.4 *Close up of coir mat*

Figure 6.5 *Bamboo flooring*

6.4 SUPPORTING MATERIALS FROM CROPS

The following materials are crop-based and can be found in some floor coverings, although they are not generally the main component. Several of these materials are also used within other products, and insulation in particular.

6.4.1 Flax

In the context of floor covering, flax occurs mainly in the form of felt for use as an underlay. However, flax is also used in the form of linen edging tapes, both for decorative wool and sisal rugs and for the ends and edges of coir and other natural floor coverings.

6.4.2 Hemp

Hemp can provide a very strong and durable backing for carpets, and is used in this way in the top range of one European manufacturer of 100 per cent pure wool carpets.

6.4.3 Cotton

Cotton is used in some carpets, generally in the supporting structure. In textile form, cotton rags are used to make the attractive, low-cost rag mats that are produced in Scandinavia and the USA. From an environmental viewpoint, most cotton production is problematic, as a high proportion of all pesticides used in the world are applied to cotton crops.

6.4.4 Plant dyes

Plant dyes are used in a very small number of natural carpets. They provide very attractive colours, but in some cases are more prone to problems of fading in strong sunlight than their chemical counterparts.

Most dyes used in floor coverings are artificial. Dyes and pigments continue to account for a third or more of all newly-developed compounds from the chemical industry. Environmental legislation, where properly enforced, has significantly reduced pollution from the manufacture of chemicals in Europe and America, but even here major leaks from chemical works periodically colour one river or another.

6.5 DESIGN AND PROCUREMENT ISSUES

The main performance issues for a floor covering are:

- colour
- wear resistance
- slip resistance
- thermal comfort
- acoustic damping
- static build-up
- ease of maintenance/cleaning
- colour fastness
- cost.

All these factors need to be considered in combination with the other materials within the space to achieve an appropriate balance. For example, if the walls are all strongly reflective of sound, it will be important for the flooring to be absorbent or the space will echo too much and be noisy. Information on each property should be available from the manufacturers or suppliers.

There is in general no simple relationship between the material and a particular performance property, as different versions of the same material will perform differently, depending on the quality of the product and the other materials it works with.

Appropriate underlays are important for achieving good thermal and acoustic performance from carpeting, as well as comfort underfoot.

Low-pollution adhesives are available for gluing down linoleum and natural carpeting (as well as timber flooring, cork, stone and terracotta tiles). These adhesives perform at least as well as synthetic adhesives.

The main materials listed above are available to order from flooring suppliers and specialist builders' merchants. Note, that as with all materials, a large order may require a longer delivery time. For the more unusual materials it will be important to agree delivery time with the supplier, as these may be imported on an occasional basis.

6.6 APPLICATION

6.6.1 Residence for the elderly, Barnet

As part of a refurbishment of an elderly person's residence new wool carpeting was chosen to replace the existing flooring. The 100 per cent natural flooring improves indoor air quality by not introducing into the building the synthetic glues and chemical treatments normally applied to wool carpeting. These synthetic treatments, often including a pesticide and stain inhibitor, release gases causing irritation to people who have a sensitive respiratory system.

By using natural floor coverings fixed to the sub-floor with a natural latex adhesive, these potential problems were avoided. The new floor coverings were manufactured to a high ecological specification and are independently tested to ensure they are free from synthetic chemicals. Their natural qualities mean that, when they come to the end of their useful life, these floor coverings can be lifted and shredded on site and then composted, avoiding disposal costs to landfill or incineration.

Figure 6.6 *Nursing home 1*   **Figure 6.7** *Nursing home 2*

Slight variations in colour may well occur in natural carpeting, even within a single roll. Raw wool and other natural carpeting materials vary in natural colour, and plant-based dyes, where used, can also vary.

7   Applications of crop-based geotextiles

7.1   WHAT ARE GEOTEXTILES USED FOR?

Geotextiles are permeable textiles or fabrics used in conjunction with soil, foundation, rock, earth or any geotechnical engineering related material, as an integral part of a man-made project (John, 1987). The use of natural textiles in construction projects dates back to ancient times, including the use of cotton fibres in earth embankments to increase stability and strength. In the 1960s, an increased supply of readily available, relatively cheap synthetic textiles (such as nylon and polypropylene) coincided with the expansion of urban and road construction, resulting in more use of such materials in civil engineering projects at that time. This expansion of geotextile use has been maintained, with growth rates in this sector estimated to be as high as 10 per cent per annum (Jagielski, 1990).

Geotextiles consist of woven, knitted or non-woven fibres, filaments, tapes and yarns. The resultant textiles/fabrics are then manufactured into sheets, mats, cells, blankets and webs, using synthetic or natural materials, or a combination of both (Table 7.1).

Geotextiles have numerous applications in construction projects. These include:

- drainage, where the geotextile may contribute to increasing local hydraulic conductivities, so enabling flow to a sub-surface drain for example

- separation, where the geotextile prevents intermixing of poor *in-situ* soils with good quality granular materials

- filtration, where transmission of fluids, but not particles of a critical size takes place through the geotextile

- ground/slope stabilisation, where the tensile strength of the geotextile enhances the inherent stability and strength of the ground, eg reinforced soil walls

- vegetation management, where the geotextile may be used to enhance or suppress vegetation growth

- soil erosion control (including shallow slope stability), where the geotextile protects the bare soil slope from erosive forces – rainfall, surface water/runoff, wind or wave action

- soil retention as part of green roof construction, although the lower layers of these need to be more robust geotextiles.

The wide range of geotextile products used in construction projects is discussed in Ingold and Miller (1988). It is estimated that around 150 million m² of these products are used per annum in North America and western Europe alone.

WHY CONSIDER CROP-BASED GEOTEXTILES?

The main benefit of crop-based geotextiles arises where the application is needed for a limited time period only, and it is then beneficial for the geotextile to bio-degrade. This might apply when plants become established to replace their role in erosion control. Clearly the main benefit of this is that there is no plastic left in the soil. In addition the embodied energy and other embodied chemical impacts will, in general, be very much lower for natural geotextiles compared with synthetic products.

Table 7.1 below compares the overall properties of synthetic and natural geotextiles. It should be clear that the materials have significantly different properties and so are suitable for different circumstances. As discussed in the following sections there are applications where the performance of natural geotextiles is better suited to the task than the synthetic options.

Table 7.1   *Geotextile properties*

Properties	Synthetic geotextiles	Natural geotextiles
Materials used	Nylon, polyester, polyethylene, polypropylene, polyamide	Jute, coir, cereal straw, wood shavings, paper, sisal, cotton waste, banana leaves, linseed straw
Fabric form	Non-woven. Woven. Knitted.	Non-woven. Woven.
Structure of products	2 or 3 dimensional. Strips. Webs. Cells. Sheets.	2 dimensional. Mats. Blankets
Stability	Inert. UV and microbiologically stable.	Light/biodegradable*
Longevity	25 years+	6 months to 10 years
Typical relative costs	5–10 units/m^2	1–3 units/m^2
Colour	Black, green	Natural

*Unless special chemical treatments to extend durability are applied.

PERFORMANCE REQUIREMENTS FOR GEOTEXTILES

To ensure effective performance of geotextiles in each of these applications, it is essential that geotextiles and their inherent properties are "fit for purpose". Geotextiles have been designed and developed for a specific end use. With this in mind, Table 7.2 shows the potential of geotextiles manufactured from crop-based materials.

Table 7.2 *Geotextile applications, property requirements and crop-based options*

Geotextile application	Properties needed	Crop-based options
Drainage	High hydraulic conductivity Durable Standards compliant	None
Separation	Specific porosity Specific permeability Puncture, burst and rip resistance Tensile strength Durable Standards compliant	Limited
Filtration	Low porosity High permeability Puncture resistance Durable Standards compliant	Limited
Ground/slope stabilisation	High tensile strength Durable* Standards compliant	**Rye, jute, coir**
Vegetation management – growth enhancement	65–75 per cent ground cover High water holding capacity Biodegradable Rich in organic matter, nutrients Temporary	**Rye, jute, coir, cereal straw, sisal, cotton waste, banana leaves, linseed straw**
Vegetation management – growth suppression	High percentage cover Durable	**Limited, although composites of natural and synthetic possible**
Soil erosion control	High percentage cover** High water holding capacity High roughness imparted to flow High drapability Temporary***	**Rye, jute, coir, cereal straw, sisal, cotton waste, banana leaves, linseed straw**

*   It has been argued that as the artificial fill/slope consolidates, then less tensile strength is required of the geotextile – opening the possibility of natural, non-permanent products to be used. It should be noted that the use of a crop-based geotextile for slope reinforcement should be considered only after detailed analysis of the slope stability after the geotextile has biodegraded, and careful consideration of the time for consolidation vs. the time for biodegradation.

**   Although this high percentage cover may suppress vegetation establishment and growth, so an optimum cover of 65–75 per cent is recommended.

***   Temporary geotextiles rely on vegetation growth in the long-term to provide slope erosion control.

7.4 PERFORMANCE COMPARISON FOR GEOTEXTILES

This section briefly compares the relative performance of different geotextile materials.

7.4.1 Cost

Natural geotextiles cost significantly less than synthetic materials. This reflects the relative cost of the raw materials used in the manufacture of the products and the costs of developing the product (many natural geotextiles were originally used for other purposes such as transport of agricultural commodities). Natural products can be treated with rot-proofing materials (eg acrylic binders, copper napthenate, coal pitch), which inevitably increases their cost and decreases their environmental friendliness (see below).

7.4.2 Quality assurance

Natural products are often inherently variable in terms of fibre, yarn and textile characteristics. The nature of synthetic fibres and yarns means they can be manufactured to very rigorous standards, which are reliable and constant. While many engineers prefer this reliability and constancy in terms of erosion control and vegetation establishment, the variation in natural geotextiles does not appear to affect performance to any significant extent.

7.4.3 Availability

An important consideration to geotextile specifiers is the availability of the product, especially when time management is critical in many construction projects (not least limiting the time a bare soil slope is exposed to the forces of erosion). In the past, there have been concerns over the supply chain for natural products sourced from outside the UK, which have to be shipped in relatively small quantities because of the limitations of bulk storage of biodegradable products (see below). But recently, organisations such as the International Jute Study Group, have researched problems in the supply chain, and reliability of supply is much less of a concern.

7.4.4 Storage

Natural products have a natural lifespan before they begin to degrade, even in storage, although contact with wet soil and direct sunlight on-site will accelerate these processes. Consequently, storage of natural products is, in general, more sensitive to degradation than synthetic products, so retailers may keep smaller stocks, which can adversely affect the supply chain.

7.4.5 Durability

Synthetic products generally have a much longer lifespan than natural products. There are few studies on the durability of erosion control and vegetation management geotextiles, but data suggest that once installed on-site, natural fibres such as jute have a lifespan of between six months and two years as has cereal straw, while coir has one to 10 years. Actual rates of degradation will depend on moisture and temperature conditions on-site, which are the very factors that affect vegetation growth. This means that conditions which encourage degradation of a natural geotextile are the same conditions that accelerate vegetation establishment and growth, and so the degradation is of reduced concern.

7.4.6 Environmental friendliness and sustainability

At a time when the landscape industry players are sensitive to environmental friendliness of products specified, natural geotextiles would appear to be preferred over synthetic products, for sustainability-related reasons. Synthetic products are made from oil-based components, which are finite.

7.4.7 Aesthetics

Geotextiles made from natural materials tend to blend into the landscape more than synthetic products, which often include carbon black to ensure resistance to breakdown by UV rays. Some synthetic manufacturers have tried to remedy this aesthetic problem, by dying their products green, but often, this artificial colouring makes the products actually more visually intrusive than those treated with carbon black.

7.4.8 Installation procedures

Typically, geotextiles are installed downslope, anchored at the top of the slope by burying around 0.2 m of the geotextile in a trench to a depth of around 0.15 m. The number of overlaps should be minimised as much as possible, with longitudinal edges and roll junctions overlapped by 100 mm. It is typical that the geotextiles are then pinned at 1 m intervals down each longitudinal edge, with a further row of pins fixed at 1 m centres down each strip.

The fixing pins used for both synthetic and natural products are usually made of galvanised wire, or occasionally small, live posts are used, which are designed to root and sprout, so protecting the slope from erosive forces even further. Further details of the approach needed should be obtained from the manufacturers.

Synthetic erosion control products are usually specified as being buried beneath the topsoil. Here, the top two centimetres of topsoil are removed from the site, the geomat is laid on the surface, and then pinned. Then seed is broadcast onto it, and then the topsoil is backfilled into the geomat. The buried geotextile offers very little surface cover at this stage, so that seeds found at the soil surface can be vulnerable to bird attack, wind blow and water erosion in the critical window before emergence. Also, the newly backfilled soil can be susceptible to detachment by rainfall and runoff.

With the natural products, the bare soil is first seeded, then the mats are laid directly on the seeded soil surface, pinned in the same manner as for the synthetic products, but with no need for topsoil removal and replacement. The seeds are protected by the surface cover afforded by the natural mat, and can benefit from the temperature and moisture conditions beneath the geotextile.

Clearly these differences in installation procedures have implications for cost of installation and technical performance.

7.4.9 Maintenance

Both natural and synthetic geotextiles require very little maintenance once installed, although the vegetation growing through them may require management (Coppin and Richards, 1990). Over time, the natural products will biodegrade and be incorporated into the top soil, with some manufacturers claiming that this adds organic matter and nutrients to the soil, so encouraging plant growth. Synthetic products are usually resistant to biological and light degradation, so can remain intact on the slope for several years. Indeed studies have shown these products to be virtually unaltered for 20 years or more.

7.4.10 Technical performance

The following sections review the technical performance of natural geotextiles, in the main applications of soil erosion control and vegetation management.

7.5 DESIGN ISSUES

7.5.1 Use of crop-based geotextiles in soil erosion control

Newly constructed slopes can be susceptible to soil erosion processes, because of:

- exposure of highly disturbed, structureless surface soil and subsoil, devoid of vegetation

- over-steepened slopes (often used to minimise land-take)

- compacted slope forming materials.

A variety of techniques can be used to reduce erosion by covering the soil. The natural geotextiles aim to prevent erosion for long enough for vegetation to develop and prevent subsequent erosion after they have rotted away.

Erosion of soil leads to subsequent difficulties in vegetation establishment, and preferential loss of nutrients and organic matter. Off-site, the eroded sediment acts as a pollutant in watercourses, adversely affecting aquatic ecosystems, and deposition of the eroded sediment reduces channel capacities and increases flooding risk.

The risks of erosion and its environmental impacts highlight the need for erosion control measures and prompts the question: what products should be specified for which conditions? This requires an assessment of the vulnerability of a construction area to soil erosion by means of site investigations. These will consider the following:

- the mean annual rainfall for the site which will reflect the influence of rainfall erosion (values can be obtained from local meteorological offices etc)

- the slope of the site as a percentage based on survey plans

- the soil texture

- the soil tilth based on surface observations

- soil structure and drainage status based on profile pit logs

- slope length based on survey plans

Studies exist of comparative trials of natural and synthetic geotextiles and their ability to control soil loss (eg Armstrong and Wall, 1992; Cancelli *et al*, 1990; Cazzuffi *et al*, 1991; Fifield and Malnor, 1990; Godfrey and McFalls, 1992; Sutherland and Ziegler, 1996; Rickson and Vella, 1992; Fifield *et al*, 1987; Reynolds, 1976; Rickson, 2000a, Hewlett *et al*, 1987).

Some products can reduce soil loss to less than 10 per cent of that from bare soil slopes (Figure 7.1). However, these studies show that geotextile performance is often site-specific: the effectiveness of the products depends on the environmental conditions of a site, ie rainfall, slope gradients, soil type. This finding implies that careful site assessment must be undertaken before any given product is specified.

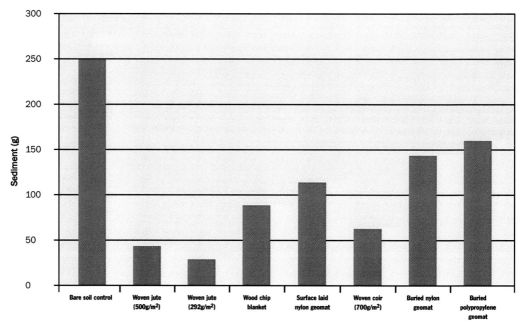

Figure 7.1 *Soil loss from slopes using different erosion control measures*

From these studies the desirable properties of an erosion control geotextile can be identified. The extent to which different geotextiles have these properties is shown in Table 7.3 below.

Table 7.3 *Salient properties of selected erosion control products*

Material	Properties							
	Cover (%)	Wet weight	Ability to pond water	Water holding capacity	Roughness imparted to flow	Availability	Ease of installation	Cost effectiveness
Woven jute (500g)	***	***	***	***	***	**	***	***
Woven jute (300g)	***	**	**	***	***	**	***	***
Woven coir	***	**	**	**	*	**	***	**
Woodchip blanket	***	**	**	**	**	**	***	**
Synthetic (buried)	*	*	*	*	*	***	*	*
Synthetic (surface)	**	*	*	*	**	***	**	*

Key: The extent to which materials have these salient properties
*   =   low
**   =   moderate
***   =   high

7.5.2   Use of crop-based geotextiles in vegetation management.

Vegetation establishment and growth

Geotextiles can be used to enhance vegetation establishment and growth on newly constructed sites by creating an erosion-free environment (see above), and by altering local microclimates, in terms of temperature and soil moisture content, which affect the rates of seedling emergence and subsequent growth. Some products assist vegetation growth by containing seed mixes, fertilisers or even pre-grown grass turf within the fabric of the geotextile itself. Many of the geotextiles used for soil erosion control are also used for vegetation establishment and growth (Table 7.4).

Table 7.4 *Examples of geotextiles used in vegetation establishment and growth*

Synthetic materials	Crop-based materials	Other natural materials
Polyethylene net – 3 layer Polyethylene net – 4 layer Liquid polymer Nylon net	Jute, coir, cereal straw, linseed straw, hay	Paper mulch, wood chips

To date, there have been a limited number of comparative trials which quantify the effectiveness of such products in performing these functions (Rickson 2000b). These trials, in differing environments, have tested the effectiveness of natural and synthetic geotextiles in terms of initial seed germination, emergence, development of vegetation and total vegetative biomass. From these trials, it is possible to identify the important properties of geotextiles used for vegetation establishment and growth (Table 7.5), which aid the selection process.

Table 7.5   *Geotextiles for vegetation establishment and growth*

Property (relative to bare soil)	Synthetic geotextiles		Natural, crop-based geotextiles	
Control surface soil erosion	+		++	
Increase germination	+		++	
Increase soil night temperatures	++		+*	
Reduce soil daytime temperatures	+		++	
Increase soil moisture content	✗		+	
Addition of nutrients	✗		+	
Addition of organic matter	✗		++	
Seedling penetration	Woven	Non-woven	Woven	Non-woven
	+	✗	+	✗

++ very strong effect
+   strong effect
✗   no effect
*   temperature increase limited by higher moisture holding capacity of the geotextiles

Vegetation suppression using geotextiles

In some applications, landscape contractors may wish to suppress vegetation growth, for example, controlling weed growth, or reducing natural ecological competition between desirable and undesirable species (Coppin & Richards, 1990). Here, the reduced solar insolation and physical barrier to emerging seedlings provided by geotextiles with high percentage cover (see above) is beneficial. This application has been called "agro-plant mulching" (CFC/IJO, 1996).

Table 7.6   *Use of geotextiles for vegetation (weed) control*

Properties	Synthetic	Natural
Barrier to seedling emergence	++	++
Reduced water infiltration	++	+
Reduced gaseous transfer	++	✗
Reduced solar insolation	++	+
Durability	++	✗

Key as for table 7.5.

The geotextile is installed on the slope, once it is cleared of any vegetation. The desired vegetation species are then planted in slots cut in the geotextile layer, which must be small enough not to allow diffuse light penetration, yet large enough to allow for growth of the stems of the selected vegetation. Depending on the planned vegetation succession, the specifier would chose whether a permanent (synthetic) or temporary (natural) geotextile would be required (Table 7.6).

7.5.3   Use of crop-based geotextiles in unpaved road stabilisation

Synthetic woven and non-woven geotextiles have been used extensively in unpaved road construction (eg haul roads, logging roads, temporary earth roads etc). Although, the amount of work on natural fibre products for this application is very limited, (Ramaswamy and Aziz, 1982) showed the beneficial effects of jute fabrics on subgrade stabilisation. The bearing capacity of a silty clay subgrade can be improved significantly when a jute fabric is placed between the applied load and the soil. Figure 7.2 shows that the same applied stress produces a much greater movement where there is no fabric within the soil (Ramaswamy and Aziz, 1983).

Effect of jute fabric on bearing capacity of subgrade soil

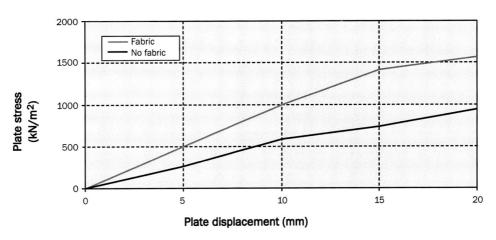

Figure 7.2   *Effect of jute fabric on bearing capacity*

It is also important to remember that the use of natural geotextiles in this application is severely limited by their inability to meet US and European Compliance Standards for properties such as durability, puncture resistance and tensile strength. So even if these products are shown to be effective in stabilising earth roads, they could not be specified as they fail to meet the minimum standards set by the industry (CFC/IJO, 1996).

7.6   APPLICATIONS

Pennine Way, Peak District National Park

For improvement and repair work on one of Britain's most famous national trails, jute and coir were used to stabilise the peat in the moor so that re-vegetation could be carried out. It is recorded that:

- both coir and jute successfully stabilised the soil

- coir is longer lasting but left unsightly strands after decomposition

- jute lasted for four years, and is seen to be more useful. However, it is more expensive and, on a large scale, is quite visually intrusive.

Source: Peak District National Park Authority (Peak District NPA)

Urquhart Castle Visitor Centre, Scotland

Figure 7.3 *Urquhart castle main view*

Figure 7.4 *The stabilised slope*

As part of the development of the Visitor Centre, new slopes were created and needed to have vegetation (grass and trees) added to blend in with the environment. Coir blankets were used to protect newly exposed soil surfaces from erosion. These mats absorbed the energies of erosion so that the soil and grass seeds were not disturbed until they were established. The coir blanket was sufficient where the soil placed was less than 200 mm deep. Where the soil was deeper, a different system was adopted.

Eden Project, Cornwall

The Eden Project in Cornwall was created within a china clay pit, and represented a substantial challenge to establish. For anything to grow, the development required extensive recreation of soil through organic generation, and the stabilisation of the many steep slopes.

In keeping with the very strong environmental commitments of the Eden Project this soil stabilisation was achieved by natural means. A natural geotextile was used to hold the soil (Figure 7.5), whilst the willow trees developed (Figure 7.6). As the trees' roots become established the geotextile will no longer be needed. The willows will be coppiced for timber, but the roots will continue to hold the soil.

Figure 7.5 *Geotextile and willow at Eden*

Figure 7.6 *Supported slope at Eden*

BP Columbian pipeline right of way restoration

Figure 7.7 *Jute matting installed for re-vegetation of pipeline*

A protection scheme was introduced to reduce the risk of soil loss on the restored pipeline right of way on steep slopes in Columbia. The erosion protection comprises jute matting laid and pinned. This was under sown with local plant species so that long term vegetation would provide a sustainable protection. The matting has been in place for some five years with no soil erosion events exceeding the target risk value.

Costs were less than £1/m² which represented a saving of some 250 per cent on alternative protection systems.

A507 Arlesey Railway Bridge – Bedfordshire County Council and the Highways Agency

Figure 7.8   *Erosion control geotextile on bridge embankment*

Figure 7.9   *Test of different erosion control products*

Figures 7.8 and 7.9 illustrate the installation of a number of erosion control products on a newly constructed embankment, next to the BR mainline, Arlesey, Bedfordshire (McKend, 1998). Natural and synthetic erosion control geotextiles were used for comparative purposes. Each product was under sown with a standard seed mix, so that long-term vegetation would provide a sustainable protection.

All products protected the slope from surface soil erosion. The non-woven products impeded vegetation growth as the seedlings could not penetrate the thick mat. Vegetation establishment was enhanced with the natural, woven products.

8   Thatch

Thatch is included briefly in this handbook because it is clearly an important use of crop-based material in the UK. However it is largely associated with houses built or refurbished in particular areas of the country that require its use for conservation reasons. This part of the industry is well supported by its craftsmen, and there is good information available. It is not likely that thatch will move into more mainstream construction, principally because of the consequences of fire. However there are a number of recent developments that have used thatch.

Data in this section are mainly taken from a number of commercial websites.

8.1   WHAT IS THATCH USED FOR?

Thatching is the use of straw or grasses as a building material. Thatch is typically used for roofing and this form of construction goes back at least as far as the Bronze Age in Britain.

A large variety of materials, such as oats, marsh reeds, broom, heather and bracken, and various grasses, have been used for thatching. Today only three materials are widely used in England – long straw, combed wheat reed and water reed.

Long straw and combed wheat reed are two methods of construction that use the same material 'wheat straw' but in a different method of application. The wheat straw is an agricultural by-product. Long straw has a distinctive "poured on" look that gives it a unique exterior appearance. Combed wheat reed has a neater, trimmed look.

Unlike wheat straw, the other form of thatching material (water reed), is not an agricultural by-product, but grown especially for the purpose. Water reed is a true reed and is regarded as a superior material, but it is slightly more expensive.

A recent development is the appearance of a product that looks like thatch, but is made from PVC. This has fire and installation benefits, but is clearly not a crop-based material and must be used with caution from an environmental view point (preferably the PVC should be recycled and recyclable).

8.2   WHY CONSIDER USING THATCH?

The benefits and reasons for using thatch construction are stated below:

- **environmental benefits** – thatch is a naturally occurring, renewable resource with a low embodied energy because energy is expended only in harvesting and transportation (generally no cultivating costs)

- **ecological and farming benefits** – the harvesting of reed maintains nationally important wildlife habitats and sustains the land usage

- **durable benefits** – thatch roof gives good durability, with a life expectancy of up to 50 years in the east of England (reduced in the wetter west to around 30 years)

- **visual requirement** – a roof made up of thatch has high standards of finish and appearance and is aesthetically sympathetic to the rural landscape

- **insulation** – thatch provides excellent levels of insulation and, expertly applied and maintained, is cool in summer and hot in winter; it has excellent insulation properties and the consequent conservation of energy represents considerable savings in fuel costs

- **operating energy benefits** – thatch is considered a "warm roof construction" and therefore, unlike a conventional tile roof, does not require ventilation, which will reduce the energy consumed during the lifecycle of a building

- **conservation and planning** – there may be planning requirements for thatch roofs, for buildings with existing thatched roofs and buildings in areas of conservation

- **deconstruction benefits** – the materials consumed in the production of thatch roofs are biodegradable.

8.3 PERFORMANCE REQUIREMENTS AND DATA

Durability/life expectancy

Good quality reed correctly installed on a roof of an appropriate design in a relatively dry area, should last a minimum of 50 years. There are many examples throughout East Anglia of roofs lasting well over 70 years and, in rare circumstances, good thatching can last more than 100 years. Climatic and micro-climatic conditions affect the durability of thatch so it is advisable to talk to an experienced local thatcher about these factors in any given location. Although there has been some concern that the quality of reed in East Anglia has declined due to poor water quality, national research has shown that this is unfounded. Good quality reed depends primarily on good reed-bed management.

Long straw	15–25 years
Combed wheat reed	25–35 years
Water reed	up to 50 years

All these materials will require re-ridging at 10–15 years intervals.

Weight

For structural calculation purposes, the weight of the roof should be taken as 34kg/m². Typical densities of thatch roofing materials:

- reed: 270 kg/m³
- straw: 240 kg/m³

Thermal resistance, R values

Typical R values for thatch roofing material:

- reed: 11.1 mK/W
- straw: 14.3 mK/W

Thermal conductivity, K values

Typical K value for thatch roofing materials:

- reed: 0.09 W/m K
- straw: 0.07 W/m K

8.4   DESIGN GUIDANCE

Pitch

Thatch should be laid and shaped so that water is carefully directed away from any points where a leak might occur, particularly junctions with chimney stacks or dormers. The steeper the pitch of the roof, the faster rainwater runs down the stems of the thatching material and off the roof. Damp does not penetrate far into the top layer of a thatched roof in good condition. Most of the thatch remains dry all the time. Unlike other roofing materials, there is no need for guttering because thatch has deep projecting eaves. This ensures that water is shed from the roof well away from the base of the walls, avoiding splash damage. However, this may require the walls to be higher than standard so the occupants will still receive a sufficient amount of natural lighting.

The pitch should be set at about 50° to facilitate efficient drainage. Dormer roofs and eaves window-roofs should be set at a pitch of at least 45°.

Thermal flow, U values

Building regulations part L (up to 2004) require a U value of 0.25 W/m^2K for the roof of a house.

> To achieve a "U" Value of 0.2 W/m^2K for thatched roofs, the following was taken from CIBSE Guide A3:
>
> The R values and K values from the performance data above gives a "U" value of 0.2 W/m^2K for the following thicknesses (thatch alone)
>
> - reed: 450 mm
> - straw: 350 mm

For thinner layers of thatch, the thermal requirement can also be satisfied by additional insulation under the roof. For example, the use of thatch batts and barrier foil underneath the thatch will give a U value of 0.16 W/m^2K.

Fire

As a thatched roof is designed to repel water, pouring water onto thatch as a fire fighting measure is ineffective. The insulating properties of thatch counteract any cooling benefit from water. Once established, a thatch fire can be sustained by oxygen diffusing from beneath the roof. Wire netting on the outside of the roof, restricted loft access and barrier boards add to the difficulties of dealing with a thatch fire.

Research has shown that a high proportion of thatch fires are chimney related, (probably as many as 90 per cent) giving a big target for strategies to reduce fire risk.

Thatch is clearly flammable and can perform poorly in fire. Various fire retardant chemical treatments are available to proof the thatch against initial ignition, but there is some concern in the industry over the effectiveness of these. It is also noted that these treatments need to be re-applied every five years. Reducing risk by good design and attention to detail is the preferred solution. This might include the use of Firewall or other materials between the reed and the roof space to protect the building from fire in the thatch. Smoke detectors should be fitted, and careful inspection of the pointing where the reed touches the chimney will reduce the risk of fire damage.

Other recommendations

- the chimney, including the pots, should terminate well above the height of the ridge for fire precautions

- there should be a smoke detector at the highest point of the roof void

- provision of a loft hatch for fire fighting purposes (minimum size = 600 mm × 900 mm). The cover to the hatch should have 30 minutes' fire resistance

- avoid recessed lighting in ceilings and external floodlights below the thatch

- avoid use of blowtorch for metal plumbing in the roof space.

Ecological assessment

Composition:	Biologically regenerative.
Manufacture:	No significant impact directly, although could consider impact of mechanised farming processes employed.
Use:	Roofing material, insulation, water repellant
Durability:	Needs to be protected against fire. Needs protection against infiltration by insects and rodents. Lifespan ranges from 15 to 50 years.
Recycling:	Can be composted.
Disposal:	No disposal issue. Biodegradable material (depending on fire retardants used).
Source:	UK and Europe.
Fire:	Performs poorly when untreated, therefore fire retardant measures are a must.
Cost:	Thatchers work in "squares" (around 10 ft by 10 ft) costing between £600 and £800 pounds per square.

8.5   APPLICATIONS

Figure 8.1   *Typical thatching straw*

Figure 8.2   *A typical older thatch building*

Figure 8.3 *The Globe Theatre*

The Globe theatre was the first public building to be thatched in London since 1666. The roof is thatched in Norfolk reed with a flush sedge wrap-over ridge. The reed was treated with fire retardants to enable its construction, and the roof also has sprinklers fitted.

Figure 8.4 *Modern thatched house*

9 Case studies

This section covers a small number of case studies to reflect the range of ways in which agriculturally-based products have been used. These support the more simply presented applications within each of the materials chapters, and seek to give an explanation of how and why the particular materials were used and, where possible, how they are performing.

9.1 ECOHOME, NOTTINGHAM

Figure 9.1 *House exterior* **Figure 9.2** *Exploded view of house*

What

Refurbishment of an existing Victorian villa.

Materials

Extensive work was carried out during the Nottingham Ecohome transformation. This comprised re-roofing, floor strengthening, new plumbing, new insulation and re-plastering. Crops were used mainly for the insulation, paints and finishes.

Claytec boards were used in the bathroom instead of plasterboard. These boards comprise clay, reed and hessian, with a clay skim. In other areas, Claytec backing plaster with Tierrafino – a clay-based material with embedded straw fibres – was used instead of traditional plaster.

In the bathroom, they used Marmoleum lino, a hard wearing sheet covering based on linseed oil and other natural ingredients. The stairs were covered with sisal woven grass carpets.

Insulation

The ground floor insulation includes the use of 100 mm Thermafleece sheepswool insulation between the floor joists.

User response

- **Claytec boards** – expensive and delicate, but users are pleased as they moderate moisture and smells and make for a very soft and clean atmosphere for the bathroom

- **Claytec plaster** and **Tierrafino** – a little fragile but comes in different colours which can be mixed to form new shades

- **Auro gloss paint** – users' favourite gloss paint, but was noted that the colours were somewhat limited

- **Green paints** – offered best colour range; the pigments, however, are petrochemical based

- **Caesein paint** – non-toxic and the colour pigments are sold separately, increasing flexibility. Labour intensive, not scratch proof and needs 4–5 coats on a bare plaster wall

- **Sisal carpet** – hard wearing but quite harsh on the feet.

For further information, check: <www.msarch.co.uk/ecohome>

The ecohome
features at a glance

1   Flat-plate solar collectors for heating water
2   Roof insulation 300/400 mm thick, made of shredded surplus newspapers
3   Roof lights with insulating (low emission) glass
4   Natural plasters – clay and lime-based
5   Super-insulated hot water tank
6   100 mm ozone-friendly drylining to front face to maintain exterior brick appearance
7   150 mm exterior wall insulation with rendered finish
8   Space-saving bath and thermostatic shower controls can save water
9   Heat-recovering fans limit ventilation heat loss
10 Environmentally friendly paints
11 Draught lobby in porch
12 Triple- and double-glazed timber windows treated with natural fungicides and stains
13 Energy-efficient appliances
14 Second-hand, natural and reclaimed furniture
15 Stripped floorboards
16 Copper rainwater goods with filter for rain harvesting
17 160 mm natural floor insulation
18 Rainwater storage for use in WCs, washing machine and outside tap
19 Low-flush WCs
20 Non-PVC wastepipes
21 Composting chamber for solid waste from WCs
22 Separator lets liquid drain off and solids into composting chamber
23 Decking from English green oak provides longevity without toxic pressure treatment
24 Organic land management utilising the principles of permaculture. Growing our own food saves on packaging and transport

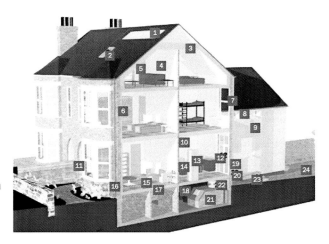

Figure 9.3 *Features of the house*

9.2 HAVERHILL HOUSES

What

Two houses constructed using a hemp and lime mix as a part of a housing development of 18 houses. The construction and occupation has been monitored closely to compare their performance with traditional housing.

Materials

The timber wall frames are infilled with a mix of hemp and lime matrix. The same matrix is used to form the slab, which is topped with a sand/lime screed. The foundations for the two houses are limecrete and a brick plinth and have approximately half the mass of those for traditional housing.

The following images show the process of building the houses, by the gradual building up of the hemp lime mixture within timber formwork.

Figure 9.4 *Timber frame and hemp lime*
Figure 9.5 *Hemp/lime and shuttering*
Figure 9.6 *Interior before finishing*

Summary of research findings

The new buildings have been extensively researched as they were built, to compare the systems to conventional building. The following conclusions were reached:

- overall structural and durability equal to those of traditional construction
- perform less well acoustically compared to traditional houses but still meet minimum requirement
- generate less condensation than traditional construction
- not much difference in the amount of construction waste generated in both forms of construction
- heating requirement is no greater than that of traditional houses – see below.

Energy

The following figures show the finished buildings, and then thermographic pictures of the houses. Thermographic pictures show the temperature of the surfaces of the building using an infra-red camera. The colours show the range of temperatures that are occurring, with the brighter orange and red colours showing the hottest areas where most heat is being lost, and the darker blue areas being colder.

Figure 9.7 *Completed hemp house*

Figure 9.8 *Completed masonry house*

Figure 9.9 *Thermographic image of hemp/lime house*

Figure 9.10 *Thermographic image of masonry house*

It is clear from these thermographic pictures that the windows and doors are an area of weakness in both buildings. It is also clear that the hemp house is generally darker than the masonry one, suggesting it is achieving a better thermal performance. The bright area at the base of the hemp house is where conventional masonry is being used, and could represent a thermal bridge.

Although not proved to be the case, it is likely that the difference between these two relates to moisture in the materials affecting the thermal performance of the insulation in the masonry construction. This is supported by the fact that although the calculated U values suggest masonry should be better than hemp/lime, usage data shows that hemp/lime uses the same energy for slightly higher temperatures. There must be a mechanism for this difference, and moisture seems the most likely.

Further information can be found from:

Final Report on the Construction of the Hemp Houses at Haverhill, Suffolk (Client Report No 209–717 Rev 1) 2002 – <www.projects.bre.co.uk/hemphomes>

Suffolk Housing Society – Home from Hemp, Summary Research Findings – <www.suffolkhousing.org>

9.3   REGIONAL ADMINISTRATION AND VISITOR FACILITIES, GLENCOE

What

This newly opened Visitor Centre in Glencoe is regarded by many as the greenest building in Scotland as it takes account of the sensitivity of the local landscape and the local population to visitor pressure. The visitor facilities comprise a group of buildings designed in the vernacular form of a Clachan (a small highland village).

Materials

Individual buildings are domestic in scale, sitting low in the landscape and using materials chosen for their environmental credentials: for example external doors and windows as well as the internal floor boards have been manufactured locally from untreated Scottish oak with internal doors made from Scottish hardwood. Heating is provided by a boiler that uses locally produced wood chips, resulting in a CO_2 neutral system. Natural materials such as recycled paper and sheep's wool were used to provide insulation. Water is conserved through low-flow taps, low-flush toilets and waterless urinals, while sewage is treated on site and clean water returned to the river.

Figure 9.11 *View 1 of interior of visitor centre*
Figure 9.12 *View 2 of interior*
Figure 9.13 *View 3 of interior*

Figure 9.14 *Exterior of the visitor centre*

For further information, check:

<www.greentourism.org.uk/Glencoe >
<www.gaiagroup.org/Architects/tourism/Glencoe>
<www.rics.org/about_us/awards/glencoe_visitor_facilities.html>

LOFT REFURBISHMENT AT LEEDS METROPOLITAN UNIVERSITY

What

Leeds Metropolitan University has recently gained the Environmental standard ISO 14001. It therefore decided to work to a "green" brief in its building programme and specified British sheep's wool insulation within a roof refurbishment of Leighton and Cavendish Hall at the Beckett Park Campus, Leeds. The buildings date from the 1920s and are one of many teaching and academic facilities on the site.

Materials

The main focus of the project was to improve access for maintenance staff to work safely within the roof space. With health and safety at the top of the agenda, specifying wool insulation over conventional materials was the natural choice. A spokesman commented: "We wanted to create a friendlier environment for our maintenance staff to work within, sheep's wool, being a natural fibre, causes no hazard to health whatsoever as it doesn't irritate the skin and so won't cause any discomfort to our staff when they are working up there." The ease of disposal through bio-degradation at the end of the useful life was also a positive feature.

Figure 9.15 *Installation of sheep's wool insulation*

As illustrated in Figure 9.15 above, the contractor installing the sheep wool insulation commented: "Wool insulation has been a pleasure to work with, as you see we didn't need any protective clothing whatsoever to install it as it doesn't cause any hazard to health being a natural fibre. Our skin isn't itchy or irritated which is a great bonus, it makes working with it so much easier."

9.5 TOLL HOUSE GARDENS, PERTH, FAIRFIELD HOUSING COOPERATIVE

What

This one- and two-storey development, completed in 2003, comprises 14 one-, two- and three-bedroom units arranged around a car-free courtyard. The development is part of a research project funded by the DTI, and aims to be allergy free.

Materials and impact

Natural materials have been specified where possible, and materials containing known allergens or triggers have been avoided. The impact of different ventilation systems is also being investigated. Initial perceptions of the buildings are that they have already helped to reduce the asthma symptoms of some tenants.

One tenant (aged 73) is quoted as saying: "I have been on asthma inhalers since I was 13. Since moving here I haven't had to use my ventolin inhaler, which is for when I am actually having an attack, a single time. It has really made an incredible difference". Ongoing assessment is being carried out as part of the research project.

Fairfield, in part thanks to the new allergy-reducing homes, has now been shortlisted for a World Habitat Award from the Building and Social Housing Foundation, an independent research institute promoting new ways of overcoming housing problems worldwide.

Specific materials used included natural paints, linseed oil based floor coverings, and natural fibre insulation as part of a breathing wall construction.

Figure 9.16 *Tollhouse Gardens*

For further information, see:

<www.communitiesscotland.gov.uk/Web/Site/Press/22_October_2003(2).asp>
<www.gaiagroup.org/Architects/housing/fairfield-tollhouse/index.html>

10 Current developments and future challenges

10.1 BIO-POLYMERS AND BIO-COMPOSITES

One of the key development areas for crop-based products is in bio-polymers and bio-composites. This section builds through the single material polymers through to the composites that combine fibre and polymer to produce higher performance combinations.

10.1.1 Polymers and bio-polymers in construction

Polymers, usually called plastics, have an enormous range of uses in construction, from light switches and fittings, through decorative mouldings to many water applications especially pipes and gutters.

Polymers are made by combining the small building blocks (called monomers) to produce the polymer with the desired properties. These properties depend on the "long-chain" nature of the polymer, and vary according to the length of the polymer and the chemical bonds within it. In recent times almost all polymers have been made from fossil oil sources, as these provide a ready supply of the building blocks needed. However there is no fundamental reason why the same products cannot be made from plant-based sources, although it may prove more expensive. In principle both thermoplastic (remeltable) and thermoset (unmeltable) composites can be produced.

Plant-based resins or bio-polymers are being actively developed by a number of the large chemical companies for example (Cargill Dow web), but it is interesting to note that others such as (Monsanto web) has left the field as it did not see that its investments would yield returns soon enough. This area is still mainly at an early stage of development, and there are no products yet available to buy.

Plant based resins include:

- poly-lactic acid (PLA) – based on processed corn starch
- cashew nut shell liquid – produced by pressing the shells of cashew nuts (which are very toxic)
- resins based on a variety of plant oils like sunflower, oil seed rape.

These polymers can be used directly as plastics, but also as adhesives and, as seen in the next section, as part of fibre-reinforced polymer composites. At present their use is very limited.

Fibre-reinforced polymer composites in construction

Before discussing plant-based materials within composites, it is worth relating these to the "conventional" products closest to these, namely fibre-reinforced polymer (FRP) composites, of which GRP (glass fibre reinforced plastic) is the best known.

In previous research for CIRIA (Cripps, 2002), the glass fibre manufacturers reported that some 31 per cent of the 1997 European market for their products was used in construction, representing 130,000 tonnes of fibre. Assuming around 50 per cent fibre in the FRP composites this suggests around 250 000 tonnes of FRP composites are being used in construction across Europe per year. This breaks down into the areas shown in Figure 10.1 below.

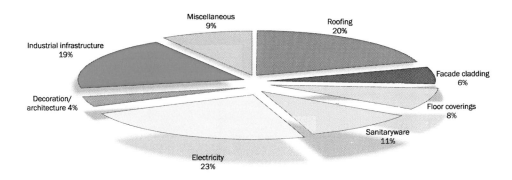

Figure 10.1 *Uses of FRP composites in construction*

Although these are significant quantities, they clearly represent only a very small proportion of the total material. Conventional FRP composites were observed to succeed in the market only in circumstances where their particular properties meant they were able to out-perform alternatives in some significant way. Generally the composite product is more expensive in first cost, so other through-life costs benefits must be possible if the product is to succeed. Possible benefits are:

- weight saving – resulting in savings in other structure or cost/time of installation
- repair/strengthening – they offer a cost-effective way to repair failing structures or those in need of strengthening; they are now the material of first choice for many conditions
- low maintenance requirements
- resists a harsh or corrosive environment
- impact/blast resistant
- can be made fire resistant
- freedom of shape/variety of appearance
- radio transparent/non-conducting/non-magnetic.

However these benefits are countered by a number of potential disadvantages:

- initial cost – often higher than alternatives
- codes and standards – there are not many
- environmental impact – not recyclable
- fire – most polymers burn, but can be treated to be very resistant
- joining – different approaches needed from conventional materials
- perceived design complexity
- finishes/aesthetics – people are not used to the look
- health and safety – different issues from those with other materials.

Bio-composites would be seen to be better on environmental impact, and also bring some benefits on health and safety. The other factors would remain to be addressed, particularly the costs.

Opportunities for bio-composites and natural fibres in construction

Like most other industries, construction is very cost-conscious and risk-averse. For a new product to succeed it must either perform better for the same (or similar) price as competitors, or give the same performance for less, and without too great a risk. It is very hard to establish a mass-market based on environmental benefits alone, although there may well be a niche market for those clients with a particular interest in environmental issues.

There has been extensive research in universities worldwide into the performance of natural fibres, and a summary of some of these data (Wambua, 2003) are given in the table below.

Table 10.1 *Data on natural fibres used in composites*

Properties	E-glass	Hemp	Jute	Coir	Sisal	Flax	Cotton
Density 1000 kg/m^3	2.55	1.48	1.46	1.25	1.33	1.4	1.51
Tensile strength (MPa)	2400	660-900	400-800	220	600-700	800-1500	400
E-Modulus (GPa)	73	70	10-30	6	38	60-80	12
Specific modulus (E/density)	29	47	17-21	5	59	26-46	8
Elongation at failure (%)	3	1.6	1.8	15-25	2-3	1.2-1.6	3-10
Moisture absorption (%)	–	8	12	10	11	7	8-25

From these it is apparent that the performance of natural fibres is as good as glass fibres on a weight-for-weight basis (the specific modulus), and can therefore give useful reinforcement of "conventional" resins. Table 10.2 below (Wambua, 2003) gives typical figures for fibres combined with polypropylene, and shows that the hemp composite is performing to an equivalent level to the glass fibre composite, and both are stronger and stiffer than the others.

Table 10.2 *Data for selected representative composites*

Fibre in PP matrix, ~30% fibre by volume	Tensile strength MPa	Flexural strength MPa
Hemp	~50	~54
Glass	~30	~60
Sisal/jute	25-30	25-35

Plant-based resins/plastics

The materials options for these were introduced in the section on polymers (section 10.1.1). However the main problem for these materials is that they are at present produced only in very small quantities and there are no established products available. There are considerable challenges to move them to a more commercial scale of production, as the research work has been focused on producing small quantities of samples for testing.

There is research continuing into the possibility of using crop-based resins with glass fibres or as the binder for wood fibre-boards. However there is more interest in the topic of the next section of crop-based resin and fibre.

Plant-based resin and fibres

By combining natural fibres with plant-derived resins a fully plant-based composite can be made. Again there is extensive research into basic properties but, as yet, there are no products available. Some people use the term eco-composite to cover materials with the lowest possible environmental impacts, and reserve bio-composites for those that are also bio-degradable.

For this area to develop there is more work needed to establish the following types of data, based on real production quantities of material:

- costs

- sensitivity to environmental attack

- inflammability

- rate of moisture absorption

- lifetime

- method and rate of bio-degradation

- opportunity for recycling

- what standards it meets

- embodied energy

- life cycle analysis (LCA).

Some typical research developments are shown below.

Figure 10.2 *Examples of boards made from bio-composite*

Figure 10.3 *Model boat made from bio-composite*

Boards

There are two classes of board products. The compressed straw boards that contain no additional adhesive, and those with adhesive applied. Both will benefit from detailed LCA work, to give a full representation of their performance.

For the compressed straw boards, at least in the UK, there is a need to establish fully why the products fell out of favour, and then to resolve if this is a good time to re-introduce them to the market. There appear to be no particular reasons why the product could not again have a place in UK construction.

The glued boards are slowly developing a place in the market. They are hampered in this by the low price of chipped waste wood at present available in the UK. As this is available in large quantities because the cost of landfill and the benefits of packaging notes, it means that waste wood is being recovered more. Another area of importance is that some users will seek to use natural resins as well as plant fibres in their boards, while at present most producers are using conventional oil based adhesives. However there are developments in the production of poly-lactic acid (PLA) glues from starch.

One aspect of both products that deserves further examination is the transport of raw material. In either case low density fibre needs to be moved to the factory to be processed into boards. It might be more effective to develop a mobile processing system that can be taken to the farm and process straw directly into boards on site. This would reduce the transport costs, but may not prove economically viable.

Construction applications

Currently none of these bio-composites is being used in construction. Because of the wide range of construction activities there are many possibilities that may emerge in the future. A number of ideas are discussed below.

Where FRP is already used

Given that many of the properties of bio-composites will be similar to those of GRP, it is reasonable to look at the applications discussed previously for GRP, and to see if these offer opportunities. A concern in this area is whether the lifetime performance of the material will be sufficient, and this remains an unknown for such new developments. This applies particularly since it is often the resistance to an aggressive environment that favours GRP over metal or timber, as is clearly the case for small boats and other marine applications.

However, in other cases, the lower cost of the natural fibre may be a significant benefit, and this may enable wider applications. This might be expected to apply to interior applications in the architectural and electrical sectors, where moisture resistance is less important. The combination of reduced cost and weight has encouraged the use of natural fibre composites in the car industry.

Environmental drivers

The major driver for interest in bio-composites is, of course, the reduction in environmental impact expected from their use, particularly in terms of fossil fuel use, but also in end of life issues. It is therefore sensible to look at areas with particularly significant concerns on environmental grounds, that could lead to good applications. This applies particularly to bio-resins since these currently are mainly more expensive than fossil fuel derived ones, and there needs to be strong enough drivers to make them worth using.

A good example of this are the concerns about the impact of PVC, and attempts by some environmental groups to see it phased out because of concerns over health effects (Greenpeace). Some projects for green clients now insist on being PVC free, and this is gradually creating demand for alternatives that could be filled by bio-composites if the price is low enough. In addition, because PVC is widely used for water and drainage applications, there is a clear need for excellent water resistance and durability. It may not be necessary to use reinforced plastics for these applications, but the fibres may reduce the cost of the product, rather than increase it, since bio-polymers are currently expensive.

Other potential applications

Other potential applications emerged during the production of this handbook, but these have not been evaluated.

- because of concerns over the performance of strawboards, the idea of layering this with higher performance materials was put forward

- if the right aesthetic performance can be achieved, then materials could be used as permanent formwork for concrete

- fibres can be added to concrete to reduce cracking

- cladding for "green" buildings – a potential new aesthetic attribute.

Challenges to overcome

There is a long way to go before any widespread take up of bio-composites. The main challenges relate to the cost of production, shifting this to larger scale, and developing products that meet the needs of consumers. Users will also be confused by some of the terminology, whether eco- or bio-composites are better, whether the resins are biodegradable or not and how much of what they are buying comes from crops. More research is required in this area before significant progress can be made.

10.2 STARCHES

Brunel University and others are working on expanded starches (Wang, 2001). These have potential applications as insulation products, but have the greatest potential in packaging, because of the rapid way in which they bio-degrade. The images below show starch foam materials, and three pictures of foam affected by rain after 0, 10 and 20 minutes.

Figure 10.4 *Expanded starch packaging*

Figure 10.5 *Block in rain 0 minutes*

Figure 10.6 *Block in rain 10 minutes*

Figure 10.7 *Block in rain 20 minutes*

Construction packaging is a significant problem on sites, and the use of starches could form a part of the solution. It will, of course, be important that the packaging does its job until the products it is protecting is ready to use. Normally products are protected with expanded foam, then cardboard and then plastic... so the foam does not itself need to be waterproof.

Given that these are designed to bio-degrade rapidly they are unlikely to have other applications in permanent construction, but there may be opportunities in temporary works or very short term buildings, eg for film sets or similar.

There is a social concern over the use of food-based starch. There may well be an excess of supply over demand in the West at present, but this is clearly not the case world-wide, and the use of these materials on a large-scale may well prove contentious.

10.3 GLUES AND SEALANTS

In the production of glues (adhesives) modern technologies have generally been found to give far greater performance than simpler, agricultural-based alternatives. This is to the point where they are not capable of doing the same job, when it is possible to consider gluing steel beams together in place of welding them.

There are therefore no current applications of crop or animal based glues in construction. However for lower strength applications like wall papering, the natural option is viable.

Historically most glues were animal or plant-based, and these have been used for many centuries for a wide range of applications. Berge, in his book *The ecology of building materials* (Berge, 2000) describes the wide range of types and applications of glues, and these include plant and animal glues.

There is considerable research into the development of resins based on plant sources of oil and starch. These were discussed in section 10.1.1 on natural bio-polymers.

10.4 STRUCTURAL BAMBOO

In the appropriate climatic zones bamboo is widely used both for permanent structure and in scaffolding. While it is clearly possible to import bamboo for similar uses in the UK, it is not expected to become established as a regular construction material. Bamboo has good strength to weight performance, but suffers from bio-degradation in humid conditions.

There is no obvious case for the widespread use of structural bamboo in the UK, and no champion trying to encourage it. Is this the way it should be? The use of bamboo in structures can be beautiful, and experience from Venezuela shows how effective it can be in earthquakes. It is certainly appropriate to use it locally for construction work in bamboo producing countries, but there is less of a case for its import to the UK.

As an example of an ambitious project using bamboo the following example explains many of the issues.

Application: Hong Kong open area theatre

This building is a temporary bamboo open air theatre constructed as part of the "One Vision Two Cities", Festival 2000 realised by the Hong Kong Institute of Contemporary Culture and the Haus der Kulturen der Welt of Berlin in collaboration. It was erected for three months in Berlin before being moved to Hong Kong.

One aim of the Festival was to demonstrate contemporary usage and artistic interpretations of this bamboo as a structural material, using traditional Chinese bamboo construction techniques.

Bamboo constructions have been used successfully for many years but are difficult to analyse and design precisely. This is recognised in Hong Kong and such structures are permitted without normal building approval so long as they are constructed by experienced bamboo masters.

The structural design method, or philosophy, needs to respect these requirements. The approach taken therefore was as follows:

- define the structural system, load paths and structural diagrams and illustrate these clearly with sketches

- adopt reasonable minimum loads for the artwork, for example do not design for roof access

- carry out relatively simple and clear calculations to establish reasonable member design forces and structural behaviour

- compare design forces with calculated bamboo pole capacities

- substantiate required connection joint capacities with simple load tests

- carry out and record a trial erection.

The bamboo jointing method was of particular interest because the German building authorities required a quite rigorous justification. Traditional bamboo joints were made by soaking strips of bamboo and lashing around the poles. The arrangement and number of lashes was decided by the bamboo master based on experience and rules of thumb.

Figure 10.8   *Bamboo structure in Hong Kong*
Figure 10.9   *Bamboo structure 2*

Further information on bamboo

A leading exponent of bamboo structures is the Venezuelan architect Simon Velez (Velez, 2000), who led the design of a pavilion at the Hannover Expo in Germany in 2000, featured on the Zeri website <www.zeri.org>.

For those who wish to learn more about bamboo, there is a dedicated organisation. The International Network for Bamboo and Rattan offers a vast amount of information, although there is a membership fee: <www.inbar.int>.

11 Conclusions and further work

Through this handbook we have seen that there is enormous potential for crop and animal-based products to help to make UK construction more sustainable at the same time bringing real tangible benefits to the UK agricultural sector.

By using the existing products now, the industry can help to bring this about. Without the take-up of what has been developed already, there will be reduced incentives for new ideas and future real opportunities may be missed. But if the current products are seen to be successful, then the possibilities for future products are more likely to come through, because investors will see there is real potential. Ideas will then become reality.

Although there are many useful products discussed within this handbook, clearly there is also a wide range of opportunities to advance the available products and to clarify and evaluate the benefits or disadvantages of their use. There are many possible areas for further work, but the following have emerged as being appropriate from this work. **Insulation**

It is clear that the use of natural materials that will biodegrade once removed from use is attractive. A more thorough life-cycle analysis (LCA) of the overall performance would be valuable in helping to weigh up the comparative performance. This would require a broad study, considering the relative impact of additives to improve fire performance and discourage pests as well as the basic materials.

A particularly important benefit claimed for natural fibre insulation relates to better performance in real conditions than competing products. This is based on the fact that all insulation materials perform less well when damp, but that different materials will be affected differently. Natural fibres absorb moisture and this should mean that there is less loss of insulation. These properties need to be tested rigorously both in the laboratory, but more importantly through trials in real houses, and comparisons with other insulation options.

Paints and finishes

As with natural insulation materials, there is a presumption that natural ingredients will be healthier than chemically processed ones. Given the tradition of the use of arsenic and lead in paints through history it is clear that this is not always the case. Therefore a thorough comparison of the benefits and impacts of natural paints is needed.

Another useful study would be to re-examine the ingredients used through history and in other cultures in paints, to search for products that could be re-introduced into current UK systems where appropriate. Taking the reverse approach, it could also be useful to examine the ingredients in current paint systems, and target efforts to find alternatives to the materials that have the largest impacts.

Floor coverings

There is clearly still a strong presence of natural fibres for carpets and floor coverings, and for linseed oil in lino. As with the other topics these could benefit from an LCA treatment to give a full assessment of the benefits of using them and whether any additional cost is justified.

One specific issue raised with floor coverings is the difficulty of obtaining backing materials and glues with the same environmental benefits as the main flooring material.

Work on alternative and less harmful treatments for wool carpet would be useful.

Geotextiles

There is already extensive literature on the options for natural fibre geotextile materials. The main challenge here appears to be in encouraging their wider use in industry. This needs to address concerns particularly that may be held by the users as to whether the materials will perform long enough.

Further specific marketing to ground works contractors and landscape experts would be valuable.

Thatch

Thatch is a well understood and widely used material in certain areas of the country. The main issues in the UK is the availability of appropriate reeds, and the scarcity of the skilled labour needed to install thatch. This suggests possible work areas relating to:

- the re-establishment of wider areas of water reeds and their efficient production
- approaches to speeding up the installation of thatch
- increasing the training opportunities for apprentice thatchers.

Reinforcement in blocks/plaster

Because of the growth of interest in the use of hemp with lime, and unfired clay blocks, there has already been some recent work in this area (Bath). A particular challenge to be addressed for wall systems relates to the increasing requirements of UK Building Regulations for thermal insulation. Although fibres in blocks or lime increase the insulation value, this is not going to be sufficient for the new 'U' values without very thick walls, and these are not likely to be acceptable. There is, therefore, a need to establish an effective solution to this challenge.

The advice provided in this handbook for the amount and type of fibre to use is based on a limited amount of source information. If these techniques are to be taken up more widely, to provide buildings with thermal mass but low embodied energy, then further studies of the performance will help to increase confidence.

Straw bales

The enthusiasts for straw are convinced that it is an effective building solution, while many remain sceptical that the challenge of damp in the UK climate is too great. At present it seems likely to remain as an option for the self builder with plenty of space and time, but limited cash resources. For this to change it might require more work into the systemisation of straw bale building, perhaps developing on the work at the University of the West of England (UWE) into panels of straw.

Bio-composites

The field of bio-composites is very research active, and there are many new ideas continually being put forward. It is likely that the answers to most questions about the performance of particular combinations of fibre and resin are known already. The problem is that there is little connection between the researchers and those who might realistically produce products for the construction industry. This disconnection needs to be addressed along with serious attempts to produce commercial products. The need to work on commercialisation, the practicality of larger scale production and cost reductions apply equally to plant oil based adhesives and resins, natural fibre composites and bio-composites.

An interesting challenge for plant oil based resins is retaining sufficient bio-degradation to allow composting or similar, while giving adequate performance in use.

106

Bamboo

While it is clear that bamboo can be used in a variety of structural applications, there are significant barriers to its likely use in the UK. It is perhaps more valuable to educate UK professionals in the appropriate use of bamboo when they are working on projects in bamboo growing areas of the world, reducing the need for expensive imported steel, for example.

A more detailed study of the potential use of bamboo in the UK might find other opportunities.

Future research

In relation to many of the areas discussed above, there is no significant backer to drive forward development and investment in research and technological development. This is probably because of the small or non-existent nature of the commercial organisations involved with these technologies. This is in contrast to the generally very large and powerful organisations that produce the products that the crop-based alternatives compete with. This lack of commercial power is a major barrier that needs to be considered further and overcome if the potential is to be realised.

Future research on crop-based products will need to look at issues relating to:

- mass production to lead to a reduction of life-cycle costs of the products

- long-term optimisation of product performance as detailed above

- understanding of technology transfer mechanisms that encourage wider commercialisation

- optimisation of relationships between construction clients and crops supply chains to encourage increased take-up and increased product innovation.

In the future, the expectation is that an ever-increasing proportion of the products used by the industry will be produced in the most renewable way possible, as products or by-products of agriculture. The first step for this to happen lies in the successful take-up of proven technologies by the mainstream construction industry.

12   References and further information sources

Chapters 1–2

Allen P, 2003. "Home-Grown Houses: The potential for large-scale production of renewable construction materials from crops grown in the UK, and the possible impact on construction, the environment and farming," MSC thesis, University of East London

Construction Products Association, 2002. <www.constprod.org.uk> (front page of website)

Defra, 2002. *The Strategy for sustainable food and farming – facing the future,* Department of Environment, Food and Rural Affairs, London

DTI, 2002. *Construction Statistics Annual: 2002 Edition,* retrieved from <www.dti.gov.uk/construction/stats/stats2002/pdf/constat2002.pdf>

Hendley, 2001, article in *National Post,* April: <www.friendlystranger.com/info/hemp_01/clearheaded.htm>. Similar information also in Hemp museum <hempmuseum.org/ROOMS/ARM%20PLASTICS.htm>

IENICA, 1999. *Report from the State of United Kingdom,* <www.ienica.net/reports/uk.pdf>

Kennedy, J; Smith, M and Wanek, C (editors), 2002. *The Art of Natural Building,* New Society Publishers, Canada, ISBN 0-865771-433-9

MAFF, 1999. *A New Direction for Agriculture,* <www.defra.gov.uk/farm/agendtwo/agendtwo.htm> (accessed July 2004)

Morton, R, 2002. *Construction UK: Introduction to the Industry,* Blackwell Science, Oxford

ODPM, 2001 <www.odpm.gov.uk/stellent/groups/odpm_planning/documents/page/odpm_plan_606333.hcsp>

The Royal Society, 1999. *Response to the House of Lords Select Committee Inquiry on Non-Food Crops,* London, ref 6/99

Smith, R A; Kersey, J R and Griffiths, P J, 2002. *The Construction Industry Mass Balance: resource use, wastes and emissions,* Viridis Report VR4, CIRIA, London

Wooley, T and Kimmins, S, 2002. *Green Building Handbook, Volume 2,* Spon Press

Chapter 3: Insulation

BSI, 1996. BS EN ISO 8990:1996, *Thermal insulation – Determination of steady-state thermal transmission properties – Calibrated hot box,* British Standards Institution, London

CIBSE, 1999. *Environmental Design Guide A,* CIBSE, London

Sarmala, 1984. *Lämmön-ja kosteudeneristys,* RIL 155-1984, ISBN 951-758-044-4

Chapter 4: Light structural walls

Amazon Nails, 2001. *Information Guide to Straw Bale Building,* Amazon Nails, <www.strawbalefutures.org.uk>

Light Earth project website, <www.lightearth.co.uk/index.htm>

Stulz, R and Mukerji, K, 1993. *Appropriate Building Materials,* 3rd edition, SKAT, IT and GATE, joint publication, Switzerland, ISBN 1-85339-225-1

Swentzell, A and Steen, B, 1994. *The Straw Bale House,* Chelsea Green Publishing Company, USA, ISBN 0930031717

Chapter 5: Paints and finishes

Wooley, T and Kimmins, S, 2002. *Green Building Handbook, Volume 2* Spon Press, ISBN 0-41925-380-7

Chapter 6: Crop-based products used in floor coverings

The Guardian, 2004. <www.guardian.co.uk/medicine/story/0,11381,1144653,00.html>

Net Doctor, 2004. <www.netdoctor.co.uk/news/index.asp?y=2004&m=2&d=10>

Chapter 7: Applications of crop-based geotextiles

Armstrong, J J and Wall, G J, 1992. "Comparative evaluation of the effectiveness of erosion control materials". In: *The environment is our future,* ProcXXII IECA Ann Conf, Nevada, pp 77–92

Cancelli, A; Monti, R and Rimoldi, P, 1990. "Comparative study of geosynthetics for erosion control". In: Den Hoedt (ed), *Geotextiles, geomembranes and related products,* Balkema, Rotterdam, pp 403–408

Cazzuffi, D; Monti, R and Rimoldi, P, 1991. "Geosynthetics subjected to different conditions of rain and runoff in erosion control application: a laboratory investigation". In: *Erosion control: a global perspective,* Proc XXII IECA Ann Conf, Florida, pp 193–208

Common Fund for Commodities/International Jute Organisation, 1996. *Technical specification and market study of potentially important jute geotextiles.* Project completion report by Silsoe College, Cranfield University (Project Executing Agency). September.

Coppin, N J and Richards, I C, 1990. *The use of vegetation in civil engineering.* B10, CIRIA and Butterworths, London, ISBN 0-40803-849-7

Fifield, J S; Malnor, L K; Richter, B and Dezman, L E, 1987. *Field testing erosion control products to control sediment and establish dryland grasses under arid conditions,* HydroDynamics Incorporated, Parker, CO

Fifield, J S and Malnor, L K, 1990. "Erosion control materials vs. a semi-arid environment, what has been learned from three years of testing?" In: *Erosion control: technology in transition,* Proc XXI IECA Ann Conf , Washington DC, pp 235–248

Godfrey, S and McFalls, J, 1992. "Field testing program for slope erosion control products, flexible channel lining products, temporary and permanent erosion control products". In: *The environment is our future*, Proc XXII IECA Ann Conf, Nevada, pp 335–39

Greenwood, J R *et al*, 2001. *Bioengineering – the Longham Wood Cutting field trial*, PR81, CIRIA, London, ISBN 0-86017-881-1

Hewlett, H W M; Boorman, L A and Bramley, M E , 1987. *Design of reinforced grass waterways*, R116, CIRIA, London, ISBN 0-86017-285-6

Ingold, T S and Miller, K S, 1988. *Geotextiles Handbook*, Thomas Telford, London

Jagielski, K, 1990. *Geotextiles: a growing market as we enter the 1990s*, Geotechnical Fabrics Report, July/August, pp 6–8

John, N W, 1987. *Geotextiles*. Blackie and Son, Glasgow

Maccaferri Case Histories Project Sheet 1
<www.maccaferri.co.uk/case_histories/pdf/EP016.pdf>

Maccaferri Case Histories Project Sheet 2
<www.maccaferri.co.uk/case_histories/pdf/EP0012.pdf>

McKend, K A, 1998. "The evaluation of geotextiles used for revegetation eighteen months after installation". MSc thesis, Cranfield University

Peak District NPA (National Park Authority)
<www.peakdistrict.org/pubs/penineway.htm> end of section 5.2

Ramaswamy, S D and Aziz, M A, 1982. *Jute fabric in road construction, Proc 2nd Int Conf Geotextiles*, Las Vegas, pp 359–363

Ramaswamy, S D and Aziz, M A, 1983. *An investigation of jute fabric as a geotextile for subgrade stabilisation*, Proc 4th Conf Road Engineering Assoc Asia and Australasia, Jakarta, vol 3, pp 145–158

Reynolds, K C, 1976. "Synthetic meshes for soil conservation use on black earths", *Soil Conservation Journal of NSW*. July, 34, pp 145–159

Rickson, R J and Vella P, 1992. "Experiments on the role of natural and synthetic geotextiles for the control of soil erosion". In: Proc *Geosintetico per le costruzioni in terra – Il controllo dell'erosione, Bologna*

Rickson, R J, 2000a. "The use of geotextiles for soil erosion control". PhD thesis, Cranfield University, Silsoe

Rickson, R J, 2000b. "The use of geotextiles for vegetation management". In: *Vegetation management in changing landscapes. Aspects of applied biology*, vol 58, 2000, pp 107–114

Sutherland, R A and Ziegler, A D, 1996. "Geotextile effectiveness in reducing interrill runoff and sediment flux". In: *Erosion Control Technology…Bringing it home*. Proc Conf XXVII, IECA. pp 393–406

Chapter 10: Current developments and future challenges

Berge B, 2000. *The Ecology of building materials,* Architectural Press, Oxford 2000, translated from the Norwegian version published 1992, ISBN 0-7506-5450-3

Cargill Dow web: <www.cargilldow.com/corporate/natureworks.asp>

Cripps, A J; Harris, B and Ibell, T, 2002. *Fibre-Reinforced polymer composites in construction,* C564, CIRIA, London, ISBN 0-86017-564-2

Greenpeace: PVC campaign information: <www.greenpeace.org/international_en/campaigns/intro?campaign_id=3988>

Inbar, website <www.inbar.int>

Jayanetti, D L and Follet, P R, 1998. *Bamboo in Construction: An Introduction,* International Network for Bamboo and Rattan (INBAR), TRADA Technology Limited, ISBN 1-900510-03-0

Koolbamboo, <www.koolbamboo.com>

Monsanto web <www.metabolix.com/publications/pressreleases/PRbiopol.html>

Mwaikambo, L Y, 2002. "Plant based resources for sustainable composites", PhD thesis, University of Bath

Velez, 2000, *Grow Your Own House,* Vitra Design Museum, Germany, ISBN 3-931936-25-2

Wambua *et al*, 2003. "Natural Fibres: can they replace glass in fibre reinforced plastics?", *Composites Science and Technology,* vol 63, pp 1259–1264

Wang, B; Song, J H and Kang, Y G, 2001. "Modelling of the Mechanical Behaviour of Biodegradable Foams-from Physical Fundamentals to Applications". In: L C Zhang (ed), *Engineering Plasticity and Impact Dynamics*, World Scientific, 2001, pp 117–134

Wang B; Song, J and Kang, Y, 2002. "On Modelling of Biodegradable Foams for Packaging Applications", *Key Engineering Materials*, vol 227, pp 253–60

Zeri, website, <www.zeri.org/pavilion/pavilion.htm>

Note: all web references were valid at the date of publication.